AN ABINGTON PUBLISHING SPECIAL REPORT

QUALITY IN RESISTANCE WELDING

An analysis of the process and its control

Ir J O Zwolsman

The author worked until February 1985 at the Centre of Production Techniques of Philips Netherlands BV at Eindhoven. After that he joined Weld-Equip BV at Helmond and retired in 1990.

ABINGTON PUBLISHING
Woodhead Publishing Ltd in association with The Welding Institute
Cambridge England

First published 1991

British Library Cataloguing in Publication Data
Zwolsman J O
Quality in resistance welding
1. Welding
I. Title II. Weerstandlassen. English
671.5213

ISBN 1 85573 033 2

CONTENTS

PREFACE

Resistance welding techniques developed by NV Philips International have found widespread application in various fields making it necessary in 1971 to form a special group at the Centre for Production Techniques. The aim of this group within its terms of reference is to provide a service and conduct fundamental research to develop resistance welding to a stage which meets modern requirements regarding productivity, consistent performance and weld reliability.

Over the years many aspects of resistance welding have been studied, resulting in a better knowledge of the process. Although there have been many reports published on individual research projects, the author thought it useful to combine this diverse knowledge in a single publication without, however, claiming to represent everything.

A study of resistance welding has led, inter alia, to an understanding that, for example, precise control of the DC welding pulse provides new possibilities unattainable with conventional equipment using pulses at 50 Hz.

The author hopes this overview will inspire welding specialists to continue further research and also provide useful information for those who work with this process.

The author is especially grateful to Ing L Schuringa and Dr Ir G Tichelaar for their much appreciated expert advice.

1 INTRODUCTION

The question 'What is a resistance weld?' has been the cause of much discussion. A useful starting point is to say that a resistance weld is produced by a combination of heating and pressure. Heating, W, is a result of current flow, i, through weld material with resistance, R, for time, t, according to Joule's law $W = \int i^2 R dt$, Fig. 1.

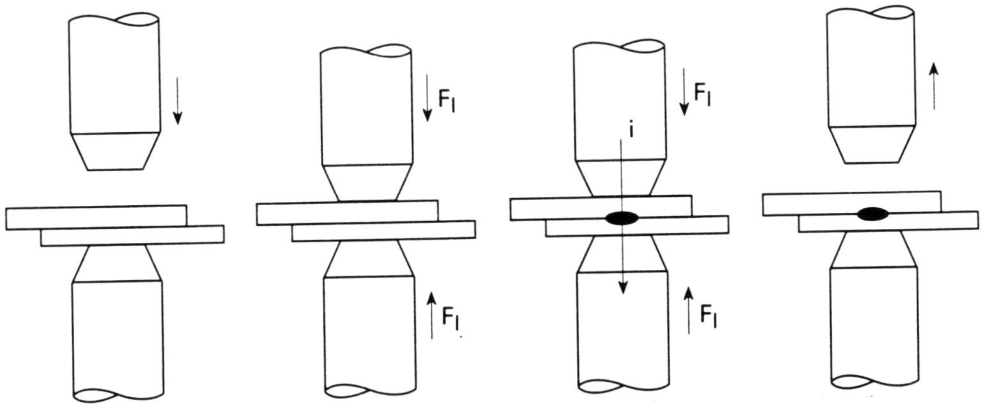

1 Resistance welding cycle.

As a result of current flow, a nugget may be formed but it is not the only factor involved in creating a strong joint; diffusion bonding and/or brazing mechanisms may occur as well.

Heating helps to disrupt surface oxide films and the parent material becomes soft or melts so that a cohesive joint is made. The electrode force, F_1, squeezes the parts to be welded into close mechanical contact throughout, breaks oxide films and often causes material deformation. Whatever the mechanisms involved, heat and pressure together determine the final result.

Processes such as laser welding, TIG welding, etc, that use heat but no mechanical force are known as fusion welding processes.

When there is no application of heat, and pressure only is present, these are known as cold forge welding processes.

The reader may feel that all the welding conditions have now been stated but, unfortunately there are so many factors which determine the final result in resistance welding that other aspects of the process need to be examined to get a good insight into what is happening during welding. Therefore in this report these aspects of resistance welding are discussed, starting with the weld set-up.

2 WELD SET-UP

2.1 Series set-up

A conventional welding set-up is shown in Fig. 2.1. Heat generation, W, and temperature, T, in the weld are determined by current flow through material of resistance, R_1, R_3, R_5, and R_7, and contact resistance, R_2, R_4 and R_6 at the spot and electrode interfaces.

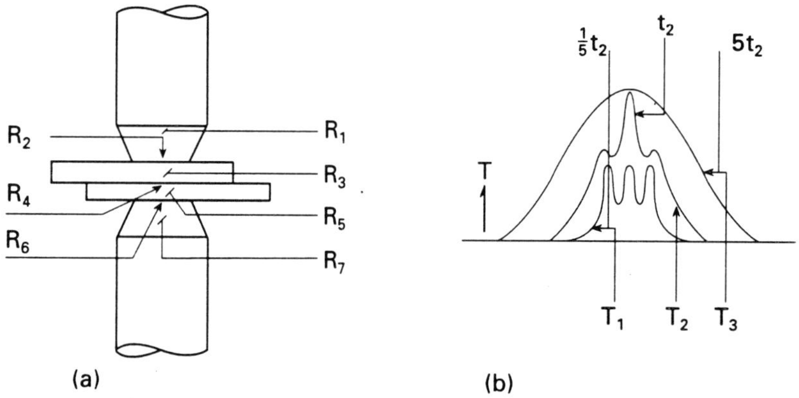

2.1 Resistance and temperature distribution during current flow: a) Electrode and workpiece; b) Temperature distribution with respect to welding time.

It is important to raise the spot interface temperature (at R_4) as much as possible, i.e. to temperature T_2 after $t = t_2$.

At the beginning for $t_1 = 1/5\ t_2$ the contact resistances are of equal importance. After a time $t_3 = 5\ t_2$ is reached, the effect of contact resistance has reduced and it is mainly the material resistance that causes an increase to $T = T_3$. In this case it might be preferable to use a constant current pulse for which heat generation is proportional to change in R, i.e.

$$W = \int i^2 R dt.$$

2.2 Parallel set-up

The situation is different using a set-up shown in Fig. 2.2.

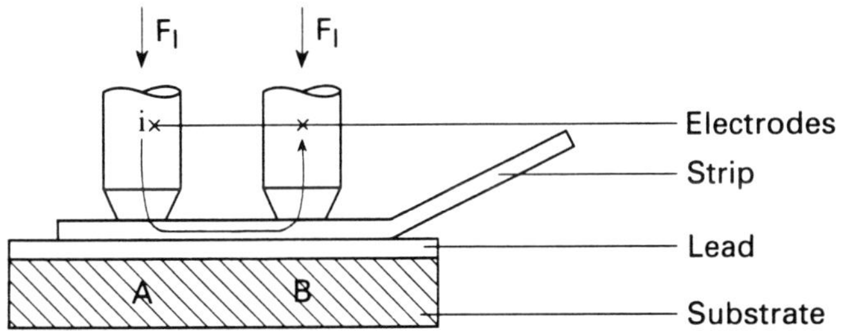

2.2 Welding with electrodes in parallel set-up.

In contrast to series operation, the two weld spots A and B in a parallel set-up are short-circuited by the strip. The resistance of the material between the electrodes increases rapidly with rise in temperature.

For constant current it is possible that the strip will melt before a weld is made. Therefore it is better in this case to weld with constant voltage, v, for which $W = \int \frac{v^2}{R} dt$.

As R increases W decreases, so that the possibility of a material explosion is minimised.

It is therefore sensible to have welding equipment available that can be adapted to constant current or constant voltage as required. This is discussed in more detail later (see chapter 13).

3 WELDABILITY AND QUALITY

3.1 Introduction

Weldability of a metal or combination of metals is determined mainly by the nature of the joint to be made, see earlier, and the quality required.

The concept of quality is not easy to express in absolute terms. Quality requirements in resistance welding depend on what the welded product will be used for. For example when a phosphor-bronze wire is welded onto a flat nickel-iron workpiece,the weld quality is excellent, but the wire material weakens considerably during transition from the unwelded to welded state.

When, e.g. welding a lacquered wire to a 0.7 mm tinned copper pin, the pin is flattened first and then folded with the wire pinched between to make a sandwich joint. During current flow a good reliable joint is formed, although the copper at the weld spot does not melt.

Thus, as already stated, quality requirements and weldability of various metals have to be determined by taking into account the use to which the welded product will be put.

3.2 Weldability

Weldability of metals can, however,be discussed in isolation from the reasons for welding them. There are metals which are difficult to weld, such as copper and others that can be welded easily, such as nickel-iron. Therefore it is necessary to consider also what is happening in the resistance weld itself.

<u>1 Material resistivity, ρ_e</u>

A resistance weld is made by concentrating welding energy in the region of maximum resistance, generally the contact resistance, R_c. But in section 6 it is shown that after the beginning of a current pulse R_c rapidly diminishes. During the short time that R_c exists, the material undergoes a transition and the temperature builds up, predominantly because of the material resistivity, ρ_e. If ρ_e is low, the above changes will hardly occur. Copper and brass plated material,aluminium and pure nickel are difficult to weld or even unweldable. On the other hand the materials with a high ρ_e such as nickel-iron, cupro-nickel, stainless steel, spring steel, fernico, etc can be welded easily.

2 Geometry

The above dealt with sheet material only, but with, e.g. copper wire, crosswire and butt welding can also be carried out, see Fig. 3.1. For sheet welding, Fig. 3.2, the steel plate may cause difficulty if the electrodes cool the joint too much, see section 10.3.

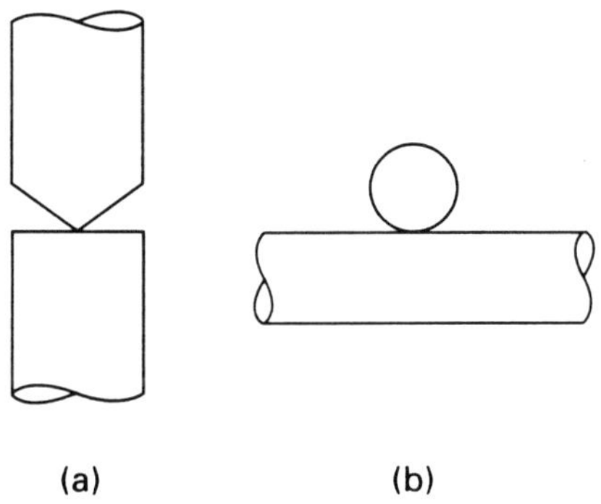

3.1 Increased R_c.

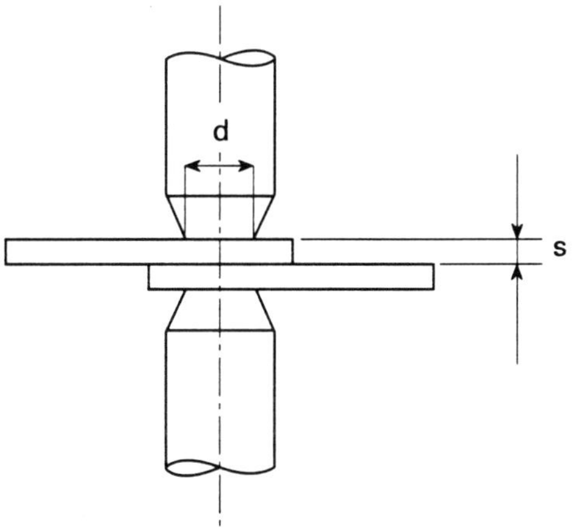

3.2 Relationship $d \leq 5\sqrt{s}$.

It was impossible using steel sheet with thickness s = 0.5 mm to concentrate enough energy via the Cr-Cu electrodes with d = 5 mm but by using the relationship $d \leq 5\sqrt{s}$ the problem was solved.

However, nickel foil of 50 μm thickness was still unweldable.

Component geometry must be taken into account to obtain the correct heat distribution.

3 Surface condition

Sometimes weldability of the two metals can be improved by adding a special resistance coating. For instance, an electroless coating of phosphor-nickel increases the resistance but has a low melting temperature of T_m = 1175 K. In addition, the material to be welded stays clean and stain free. Also, copper parts with a tin coating are more easily welded. An example is welding of tinned copper wire or strip on aluminium. Plain copper cannot be successfully welded to aluminium using resistance welding but nickel foil is weldable by adding a nickel-iron wire between the foils.

Sometimes specially deposited coatings, as used for TV tube shadow-masks, can lead to splashing and excessive electrode wear. Also, in the case of galvanized parts, electrode wear increases rapidly.

4 Projection welding

When the heat developed in the electrodes is too great, projections formed in the material to be welded are used to overcome this problem, Fig. 3.3. The material to be welded thereby concentrates the welding energy where it is needed and the electrodes can be finished flat. Projection welding is normally used for galvanized metals.

When two different materials are to be welded, projections are formed on the thinner sheet with the greater hardness and/or the higher electrical resistance. If the welding time, t_1, for two sheets with the same thickness = s_1, then welding time, t_2, for sheets with different thicknesses = s_2/s_1t_1, Fig.3.4. An example of this application is the welding of mounting brackets on transformer armatures.

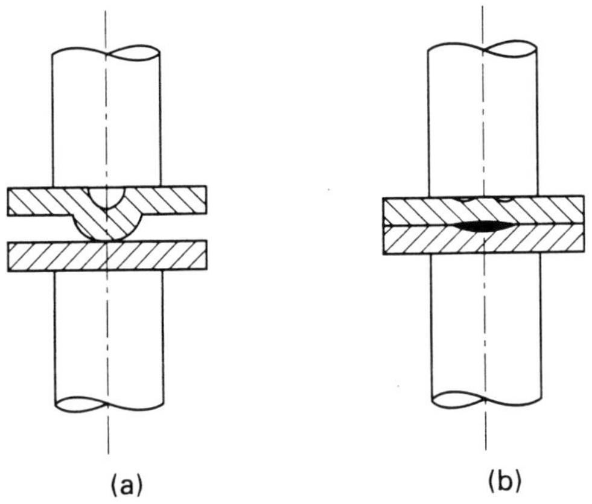

3.3 Projection welding.

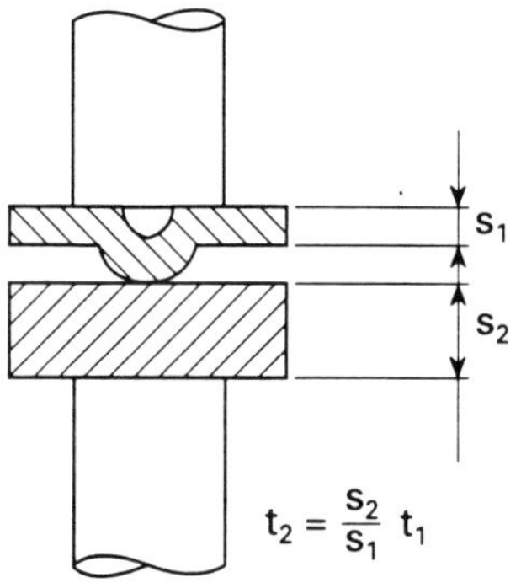

3.4 Influence of sheet thickness on welding time.

5 Direct current

When welding with direct current, attention must be paid to the correct current direction, see section 12.3.

6 Metallurgy

When attempting to join metals with a low ρ_e and high λ a satisfactory weld does not occur. It is necessary to ensure that the crystal structure of the metals, crystal sizes and coefficients of expansion are compatible, to minimise thermal stresses and prevent weld cracking.

Sometimes it is desirable to plate the parts to be welded with copper, silver, or gold. The surface is compositionally more stable and the coating, usually a few μm in thickness, improves adhesion between the two metals as indicated previously. Moreover, it is important to point out that steel which retains residual mechanical tension tends to crack when it comes into contact with copper or brass in the liquid form when it penetrates the crystal structure.

Table 3.1

Composition	Symbol	Electrical resistivity ρ_e Ωm X 10^{-8}	Melting temperature, T_m K	Thermal conductivity, λ $Wm^{-1}K^{-1}$	Specific heat, $kJkg^{-1}K^{-1}$	Melting heat, C_m $kJkg^{-1}$	Linear expansion 273-373 K X $10^{-6}K^{-1}$
Aluminium	Al	2,7	933	238,5	0,917	375	23,5
Cadmium	Cd	7,4	594	92,1	0,234	54	31
Chromium	Cr	13,0	2123	69,0	0,460	134	6,5
Cobalt	Co	6,2	1765	69,7	0,427	242	12,5
Copper	Cu	2,2	1356	382,0	0,385	172	17,0
Gold	Au	1,7	1336	300,0	0,13	67	14,1
Iron	Fe	9,7	1810	159,0	0,456	205	12,1
Lead	Pb	20,6	600	34,2	0,13	24	29,0
Magnesium	Mg	3,9	923	167,4	1,038	193	26,0
Molybdenum	Mo	6,0	2873	135,0	0,260	--	5,1
Nickel	Ni	6,8	1726	52,3	0,46	234	12,3
Palladium	Pd	10,8	1825	71,0	0,247	151	11,0
Platinum	Pt	10,6	2042	70,0	0,130	113	9,0
Silver	Ag	1,6	1233	418,5	0,226	109	19,1
Tantalum	Ta	13,5	3253	73,0	0,140	--	6,5
Tin	Sn	12,8	506	63,5	0,251	59	23,5
Tungsten	W	6,4	3623	159,0	0,150	251	4,5
Vanadium	V	26,0	2133	293,0	0,502	--	8,0
Zinc	Zn	5,9	693	106,0	0,393	117	31

Nickel-Iron	42Ni 58Fe	50,0	1698	147,7
Fernico	29Ni 54Fe 17Co	70,0	1673	220,1
Nickel-Chromium	73Ni 20Cr AlFe	133,0	1673	131,8
Constantan	43Ni 57Cu	49,0	1583	55,2
Brass 58	58Cu 42Zn 0,2Sn 2Pb	6,7	1173-1353	108,8
63	63Cu 35Zn 0,2Sn 0,5Pb	7,0	1178-1193	108,8
85	85Cu 15Zn	4,7	1283-1298	154,8
Phosphor Bronze	93Cu 7Sn	16,0	1183-1309	155,0
Steel	0,2C 0,45Mn 0,25Si Fe	10,5	1773	54,4
Stainless Steel	0,2C 18Cr 8Ni Fe	27,6	1688	20,9

Cross sectional examination of the weld provides much information, helping towards achieving the required heat distribution. As a rule, all factors which determine the heat distribution in a weld, see also section 10, must be set so that during welding the working temperature, and for a fusion weld the melting temperature, of both metals is reached at the same time. 'Working temperature' is the temperature at which the metal can be deformed mechanically without increasing its work hardening properties and without causing residual stresses. The working temperature is approximately 80% of the melting temperature. Table 3.1 gives properties of various metals.

3.3 Quality

1 Weld research

Although quality is a relative concept, general rules can be applied:

- A weld can be shown to be crack-free by cross sectional examination;
- For welding sheet to sheet the peel test is used, Fig. 3.5; if a hole is pulled in one of the sheets the weld is good.
- The torsion test of Fig. 3.6 imposes greater demands on weld strength regarding pulling of holes than the peel test. In the case of two welds of the same quality, the peel test pulls a hole more quickly than the torsion test.

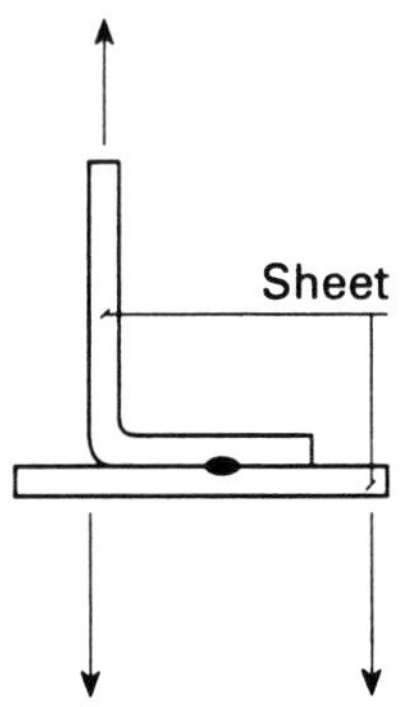

3.5 Peel test.

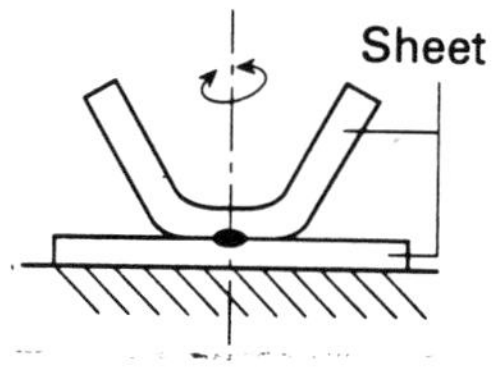

3.6 Torsion test.

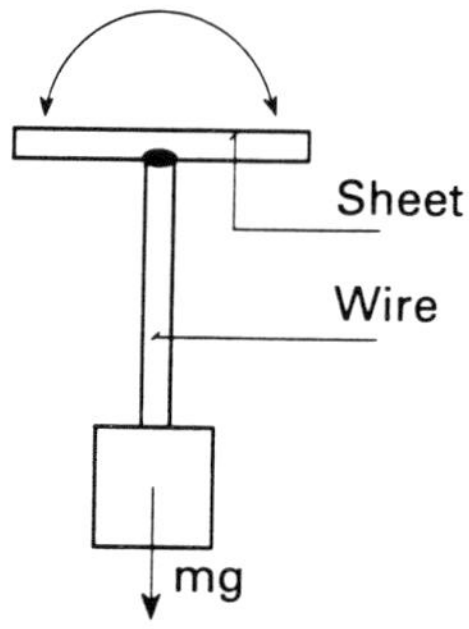

3.7 Bending test.

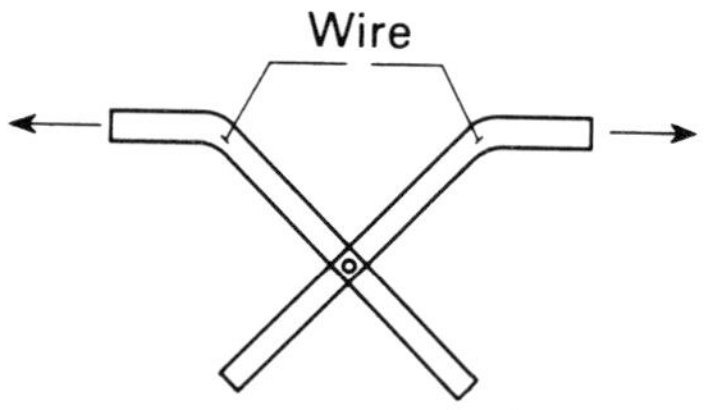

3.8 Wire tension test.

- The bending tension test is suitable for a wire to plate joint, Fig. 3.7. After suspending a weight, m, which exerts a force, mg, on the wire the plate is rotated through a number of quarter turns with respect to the wire. Both wire and weld must withstand, e.g. four quarter turns. A simple wire tension test is shown in Fig. 3.8. which yields good information about weld strength.

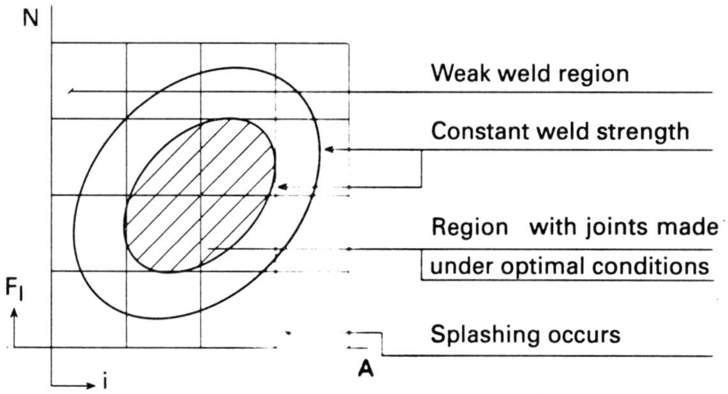

3.9 F_l/i relationship.

2 Optimising welding conditions

Ascertaining how to adjust a welding machine for good weld quality, can be facilitated by setting up an F_l/i diagram, where F = force and i = current, with F_l values on the y-axis and i, on the x-axis, Fig. 3.9. Weld strength, F_s, is used as the criterion for comparison.

A working area for a certain time, t, is created, whereby welds of a certain quality can be made. At a glance the tolerances can be seen within which the settings may vary.

4 ELECTRICAL CIRCUIT

4.1 Introduction

Once a welding machine has been adjusted to give the correct amount of energy,it is the resultant weld which receives attention from then on, in particular, whether that weld meets the requirements placed upon it. What happens when adjusting the machine is not, however, precisely known. This section investigates how a welding machine is set up and is restricted to considering a resistance welding machine working conventionally, Fig. 4.1.

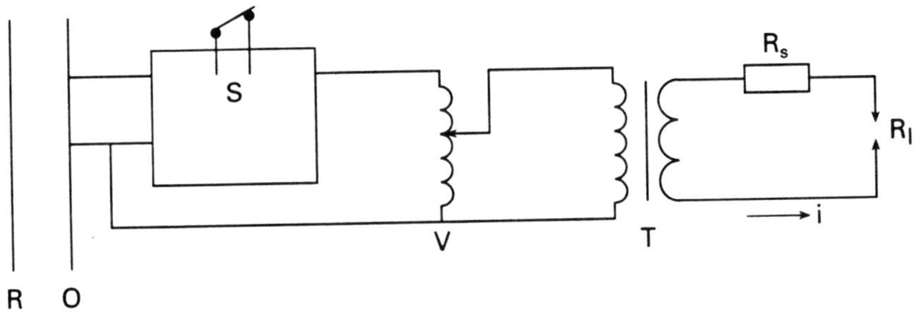

4.1 Transformer-time switch method; R=phase 220 V, O=zero 0 V.

The auto-transformer, V, continuously regulates the primary voltage of the welding transformer, T, and hence the welding current amplitude.

A normal auto-transformer is suitable for pulses of $i_s \leq 5000$ A. For $i_s > 5000$ A current adjustment via phase shift is used.

4.2 e-i characteristics

The transformer's response when current and voltage have been adjusted is shown as the e-i-characteristic, Fig. 4.2. This characteristic is easily determined by calculating short circuit current, i_o, and the open circuit voltage, e_o.

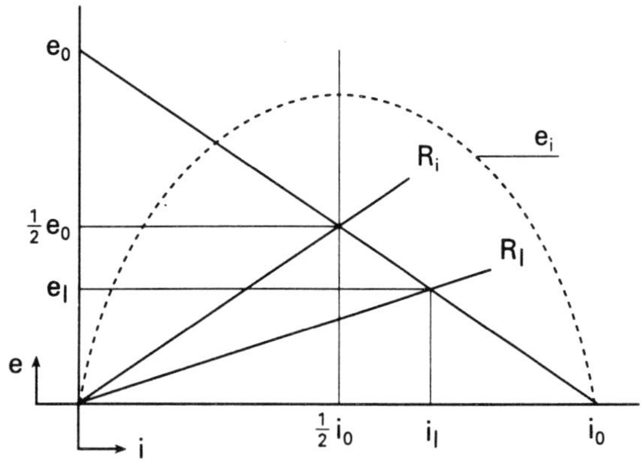

4.2 e-i characteristics.

$$R_i = \frac{e_o}{i_o}$$

The internal resistance, $R_i = e_o/i_o$, is the total circuit resistance minus R_l, the weld resistance. When the welding transformer secondary winding is loaded with a resistance R_l, the current is i_l and the voltage across R_l is e_l. The transformer supplies maximum power of $P = \frac{e_o\ i_o}{4}$ watts at $i = \frac{1}{2}i_o$ and $e = \frac{1}{2}e_o$.

Figure 4.2 shows that when R_i equals R_l variation in R_l, the load resistance, can occur but welding energy remains constant. The aim is to regulate the welding transformer to R_l which in turn meets the welding requirements.

4.3 Transformer adjustment

The e-i characteristics can be adjusted in two ways.

1 Resistance adjustment, which changes the circuit resistance, e.g. with the R_s of Fig. 4.1, R_i of the welding transformer can be changed, Fig. 4.3.

2 Voltage control. By adjusting transformer primary voltage, e.g.

with a variac (variable transformer), the e-i characteristics can be changed, Fig. 4.4.

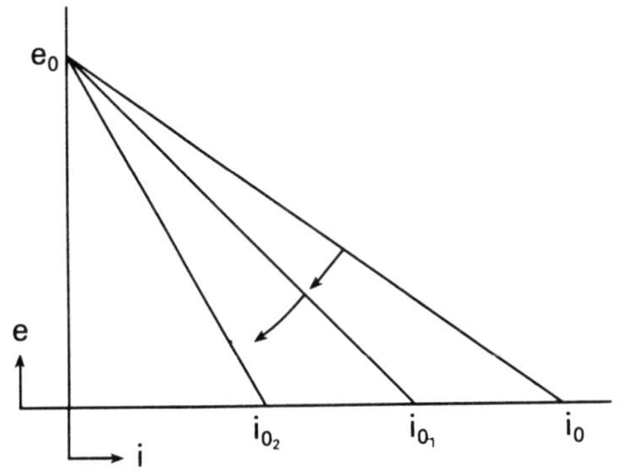

4.3 Resistance adjustment.
e_o = Constant
R_i = Variable

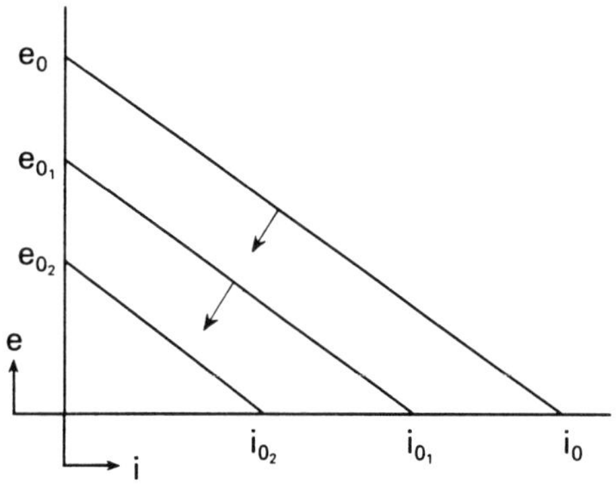

4.4 Voltage adjustment.
e_o = Variable
R_i = Constant

<u>Example</u>

Suppose that the welding transformer with an inherent output characteristic K, Fig. 4.5, is required to produce a weld with $i = i_1$ and $e = e_1$, see point L, with a characteristic K_2.

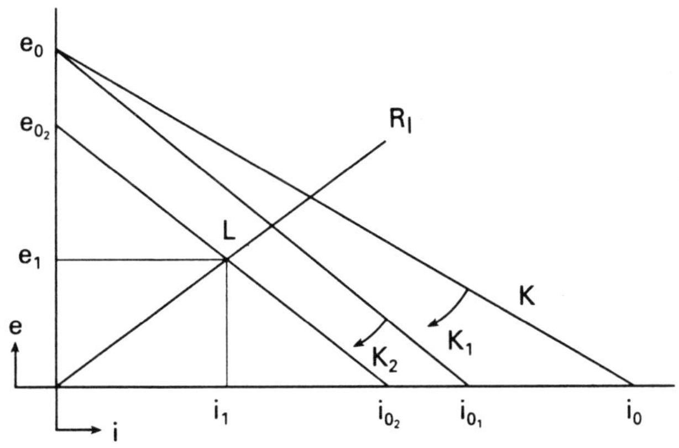

4.5 Welding machine adjustment.

It is necessary to add sufficient resistance to the welding circuit so that the gradient of K changes to K_1, which is parallel to K_2. Hereafter, decrease the primary voltage of the welding transformer so that K_1 becomes coincident with K_2. The welding machine is thereby ready to make weld L.

4.4 Which characteristic is best?

Which e-i characteristic to use depends on the welding requirements. During welding, three main constants are distinguishable.

1 Welding power or welding energy (for t = constant);

2 Welding voltage;

3 Welding current.

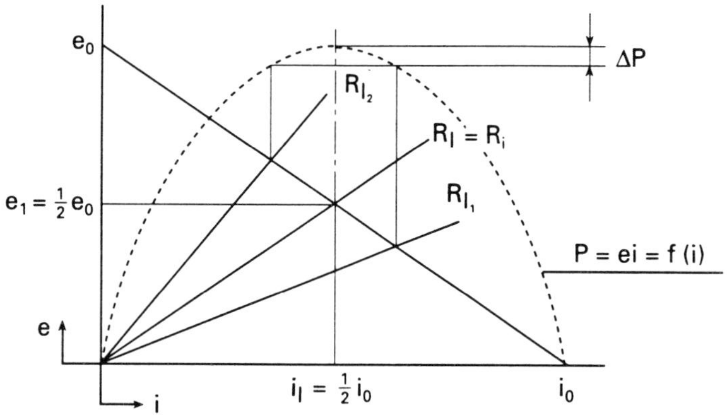

4.6 Adjustment for constant power, $i_l = \frac{1}{2} i_o$.

1 Constant power

When transformer characteristics have been adjusted so that $R_i = R_l$, then $i_l = \frac{1}{2} i_o$ and $e_l = \frac{1}{2} e_o$. If welding resistance, R_l changes, welding power change will be minimal, P in Fig. 4.6. Welding occurs at maximum power output from the transformer. If weld monitoring is needed, both voltage and current pulses should be controlled.

2 Constant voltage

If voltage has to be held constant, the transformer slope or load line characteristic will have to be as near horizontal as possible. This is called a 'hard' transformer, Fig. 4.7. In this case current monitoring is sufficient.

3 Constant current

If constant current pulsing is required, a special resistance is added to the welding circuit so that the characteristic slope is steep. This is called a 'weak' transformer, Fig. 4.8. Voltage monitoring is used with this.

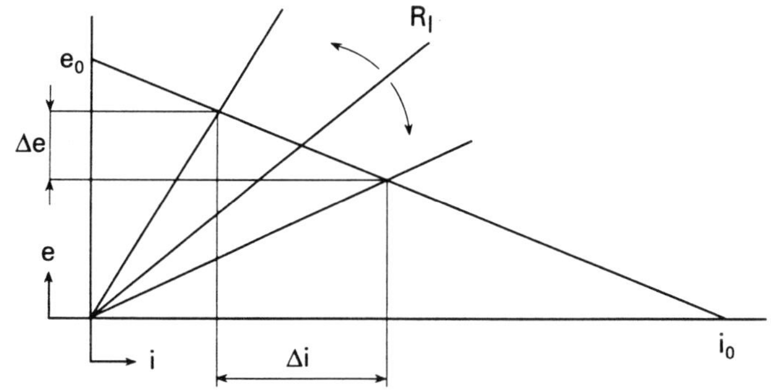

4.7 'Hard' transformer-current monitoring.

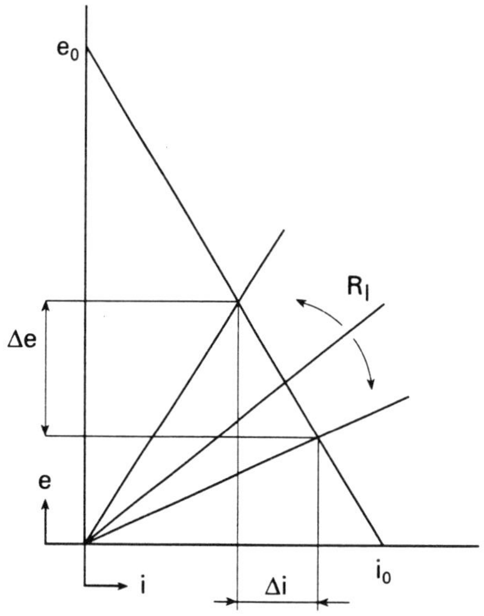

4.8 'Weak'transformer-voltage monitoring.

4.5 Applications

1 A study of sandwich welds for reed relays, Fig. 4.9, raised the question whether it was necessary to weld with a constant current pulse, i_1, or with constant energy, $W = i^2Rt = eit$.

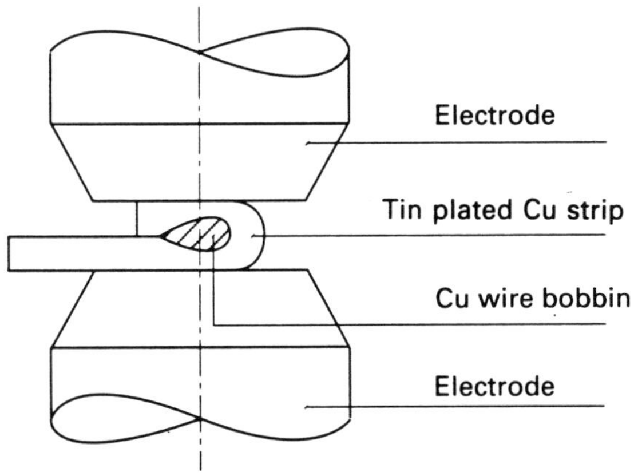

4.9 Sandwich welding of reed relays.

If the welding electrodes are dirty, heat will concentrate in an unpredictable way. It is therefore better to choose constant welding energy to be independent of variation in weld resistance. W = eit = constant, thus condition R_i = R_l was chosen.

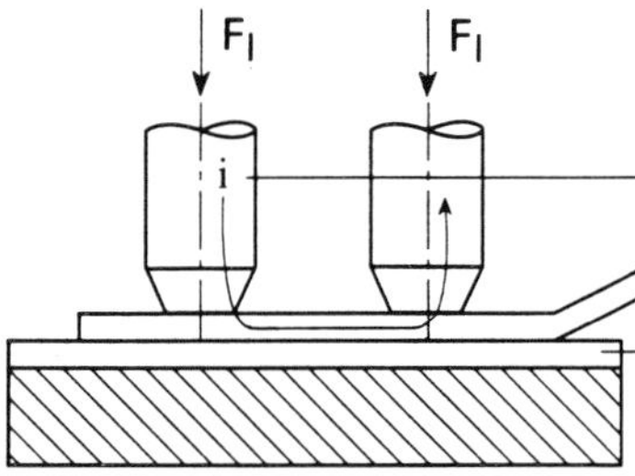

4.10 Welding of gaps between printed circuit tracks.

2 When bridging gaps in copper tracks on printed circuit boards using two parallel electrodes, Fig. 4.10, there is a possibility that a constant current pulse will cause the strip between the electrodes to become overheated and blister away from the board. The strip resistance increases with temperature and heat concentration rises proportionally. Alternatively, if the welding transformer has a gently sloping characteristic and e is constant for increase in resistance, heat produced decreases according to $W = \frac{e^2}{r} \times t$.

Thus overheating of the strip is prevented.

3 Constant current. Constant current is often preferred irrespective of variation in resistance. This requires a welding transformer with steeply sloping characteristics.

5 WELDING TRANSFORMER

5.1 Transformer capacity

A transformer's output is determined by its e-i characteristic, but a continuously loaded transformer may overheat, so a duty cycle of less than 100% is used to protect it.

$$\text{Duty cycle} = \frac{t_1}{t_1 + t_p} \quad \ldots [5.1]$$

Where t_1 = welding time,

t_p = pause time.

Normally a duty cycle of 50% is allowed for maximum capacity of $P_{max} = \frac{e_o \, i_o}{4}$, where P_{max} represents $i_1 = \frac{1}{2} i_o$ and $e_1 = \frac{1}{2} e_o$.

A current pulse $i_1 > \frac{1}{2} i_o$ requires a duty cycle < 50%, and a current pulse $i_1 < \frac{1}{2} i_o$ corresponds to a duty cycle > 50%. It is, however, impossible for a transformer to supply a current > i_o. According to the equation $i_1 = \frac{1}{2} i_o \sqrt{\frac{0.5}{\text{duty cycle}}}$ ... [5.2]

for $i_1 = i_o$ the duty cycle is 12.5% and for $i_1 = 0.35\ i_o$ it is 100%. Thus a graph of i_1 / duty cycle, Fig. 5.1, can be drawn. The manufacturer usually provides the current versus duty cycle for a transformer. Various construction factors such as air or water cooling will cause the graph to vary from the one shown.

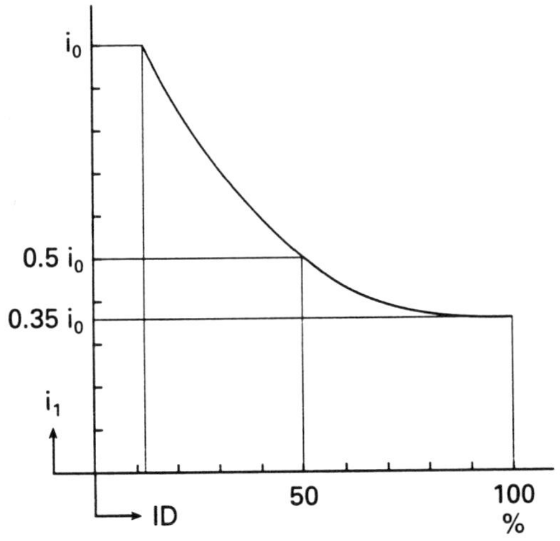

5.1 i_1 as a function of duty cycle.

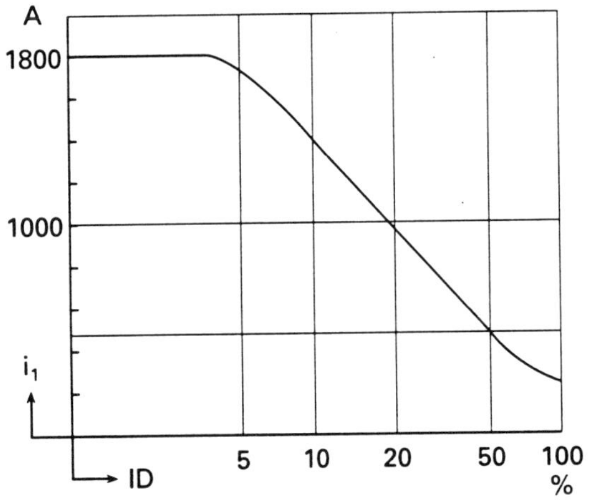

5.2 i_1 versus current for a DC machine.

For example the supplier of a DC power supply normally provides an i/duty cycle graph as in Fig. 5.2 with significant values as follows:

For $i = i_o = 1800$ A, duty cycle = 5% = 50 ms/s;

For $i = 0.56\ i_o = 1000$ A, duty cycle = 20% = 200 ms/s.

To calculate other values assume a duty cycle of 20%.

$i_1^2 \times \text{duty cycle} = 1000^2 \times \frac{20}{100} = 2 \times 10^5$ see [5.2].

For $i_1 = 1400$ A, duty cycle = $\frac{20}{196} \approx 10\%$ = 100 ms/s;

For $i_1 = 1200$ A, duty cycle = $\frac{20}{144} \approx 14\%$ = 140 ms/s;

For $i_1 = 800$ A, duty cycle = $\frac{20}{64} \approx 31\%$ = 310 ms/s, etc.

It is important to pay attention to the transformer graph and equation [5.2] from which independent values of i and duty cycle can be reasonably accurately derived.

5.2 Cos φ

Until now nothing has been said about the effect of transformer resistance when current is flowing, except for the internal resistance, R_i.

Apart from the ohmic resistance of the secondary circuit the transformer also presents a resistance to the alternating current flow. The transformer opposes these changes by developing reactance, R_s, which lags behind the ohmic resistance, X_s. Together they form the impedance, Z_s, Fig. 5.3.

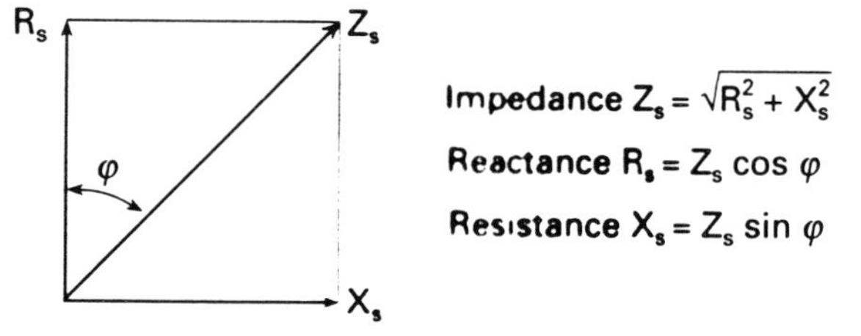

5.3 Transformer vector diagram.
Cos φ is a component of reactance.

Everything said about the welding transformer concerns the area directly around the weld. In this area there is no phase shift between current and voltage, i.e. impedance = resistance.

6 CONTACT RESISTANCE

6.1 Introduction

Resistance welding is a process influenced by electrical contact resistance, R_C, at the surfaces of the parts to be welded. When current flows the heat developed, $W = i^2 R_C t$, is concentrated at this resistance which is also called constriction resistance.

Contact resistance is indispensable in creating a resistance weld. There is, however, the strange effect that as soon as R_C operates, it disappears. As soon as the material softens through rise in temperature the contact resistance is reduced to zero. During its brief existence R_C has to be effective. In this section a description is given of how R_C functions during welding.

6.2 Holm's model

When two metal parts are placed in electrical contact current flows through the mating faces, distributed over the whole contact area, Fig. 6.1. A single mating interface is shown enlarged in Fig. 6.2.

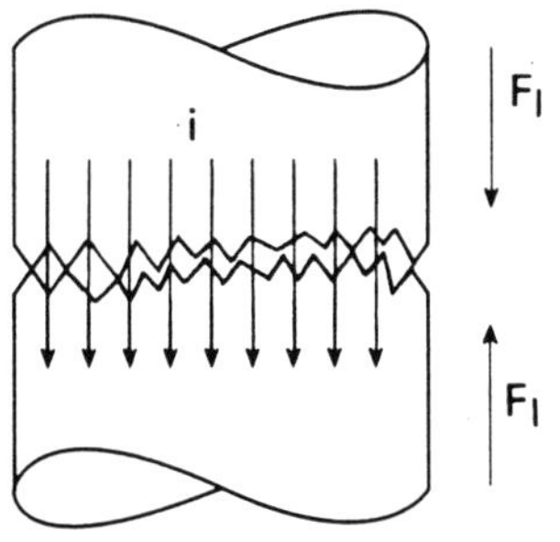

6.1 Spreaded contact places.

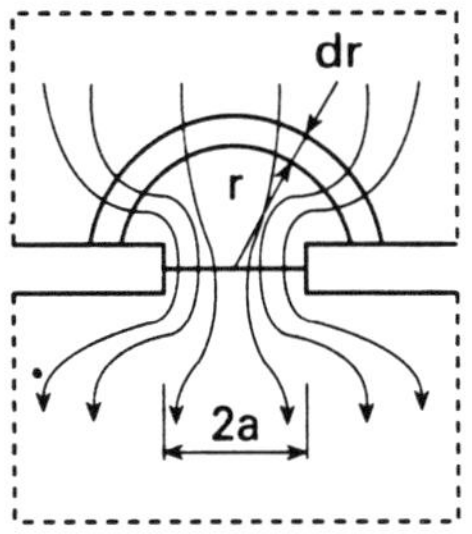

6.2 A single mating interface.

Current flow is constricted through the contact area considered to be conducting. Resistance at distance r for the small conducting hemisphere with radius a is, according to Holm,[1]

$$dR = \frac{\rho_e dr}{2\pi r^2} \qquad \ldots [6.1]$$

Resistance R_c to one side is

$$R_c = \int_a^{\infty} \frac{\rho_e}{2\pi r^2} dr \qquad \ldots [6.2]$$

$$R_c = -\left[\frac{\rho_e}{2\pi r^2}\right]_a^{\infty} = \frac{\rho_e}{2\pi a}$$

On both sides $R_c = \dfrac{\rho_e}{\pi a}$ . . . [6.3]

Current distribution in the small conducting hemisphere of radius a is not regular so that the real R_c is slightly higher than given in [6.3].

i.e. $R_c = \dfrac{\rho_e}{2a}$ . . . [6.4]

With this R_c model the material resistance ρ_e is important. Note that it is to the weld material immediately around the contact area that this applies. From this point of view it is obvious that welding copper is more difficult than welding nickel-steel.

6.3 Static model

With the R_c of [6.4] satisfactory practical results cannot be obtained directly, therefore consider three simplifications;

1 Assume that the contact area is equal to the bearing surface and is current conducting. This is not essential but is acceptable because a contact area must be deformed <u>before</u> its final state is reached, after applying the welding force.

2 The sum of all contact areas is represented by an area with radius a.

3 Surface tension of the contact area is given by

$\sigma_o = \dfrac{F_1}{\pi a^2} = \dfrac{1}{3} H_b$, where H_b = Brinell hardness. . . . [6.5]

Equations [6.4] and [6.5] represent

$$R_c = \frac{\rho_e}{2}\sqrt{\frac{\pi H_b}{3F_l}} \quad \ldots [6.6]$$

For room temperature this is

$$R_k = \frac{\rho_k}{2}\sqrt{\frac{\pi H_b}{3F_l}} \quad \ldots [6.7]$$

This equation for R_c was tested with Cu, MnNi and Mo wires, being taken care to clean thoroughly all material used. Moreover, the following need to be observed:

- Equation [6.7] shows that the contact resistance depends on material resistance; When two different materials are welded assume R_c to consist of R_{c1} with ρ_{e1}, of the one material and R_{c2} with ρ_{e2} of the other.
- R_c is dependent on the hardness of the softer material and welding force F_l.
- R_c is not dependent on the form or size of the tangent plane. For instance, both values of R_c in the equations relating to Fig. 6.3. are equal.

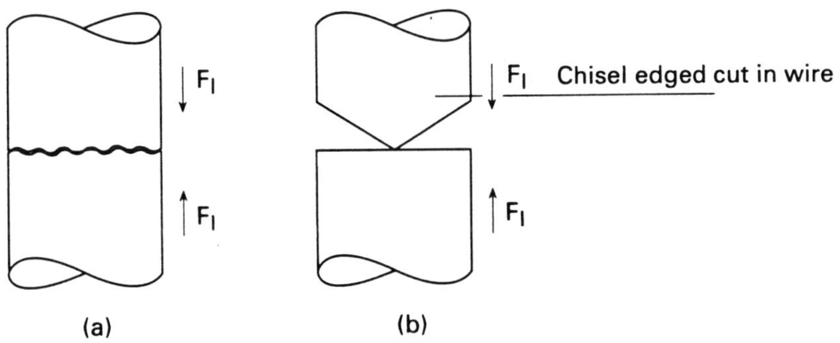

6.3 $R_{ca} = R_{cb}$ with F_l, ρ_e and H_b.

The rise in temperature when current flows is different in these two cases, as current density in Fig. 6.3b) is much greater than that in Fig. 6.3a).

- If the material is contaminated, assume a higher ρ_e than for the parent materials.

- The static model is, however, of limited use as it can be replaced by actually measuring resistance at room temperature, $T_k \approx 300$ K.

6.4 Dynamic model

It is how R_c behaves with temperature that is of significant interest.

Assume that:

1 $\rho_e = \rho_k\ T/T_k$ . . . [6.8]

Where $\rho_k = \rho_e$ at room temperature

T = temperature, K

T_k = room temperature, K

This linear dependence does not apply to all materials, however. For instance, ρ_e for Al 99.5 suddenly decreases at the melting temperature, but before this temperature is reached, $R_c = 0$. When the material is insensitive to temperature, e.g. NiCu 45/55,then equation [6.8] is not relevant.

2 Material hardness is also dependent on temperature.

$$H_b = H_k \left(\frac{T_v - T}{T_v - T_k}\right) \quad . . . [6.9]$$

Figure 6.4 shows this relationship.

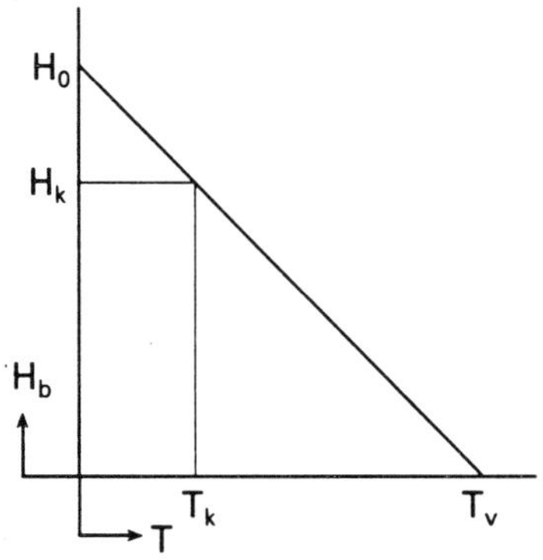

6.4 Hardness versus temperature.

In equation [6.9],

H_b = Brinell hardness

H_k = Brinell hardness at room temperature

T_k = room temperature, K

T_v = softening temperature, K

Equations [6.6] and [6.9] give

$$R_c = \frac{\rho_k}{2} \sqrt{\frac{\pi}{3}\frac{H_k}{F_l} \times \frac{T_v - T}{T_v - T_k}} \left(\frac{T}{T_k}\right) \qquad \ldots [6.10]$$

or $$R_c = C\sqrt{T_v - T}\; T \qquad \ldots [6.11]$$

where $$C = \frac{\rho_k}{2(300)} \sqrt{\frac{\pi}{3\;F_l} \times \frac{H_k}{(T_v - 300)}} \qquad \ldots [6.12]$$

C can also be evaluated by assuming $T = T_k = 300K$

$$C = \frac{R_k}{\sqrt{T_v - 300}\;(300)} \qquad \ldots [6.13]$$

When the temperature coefficient of the material is = 0, then in equation [6.10] $T/T_k = 1$ and C in [6.13] is 300 times greater.

6.5 The R_c function

The main point of giving these examples is to profile a better understanding of the factors which influence R_c and how R_c in turn reacts upon them. Moreover, equation [6.11] can be used in calculations for thermal management to suit the PHILPAC computer program.

Equation [6.11] with temperature T in K is represented in Fig. 6.5. How quickly T_v is reached depends on the upslope and amplitude of the current pulse and other factors. Figure 6.6 shows some examples where R_{c1} and R_{c2} were measured at time t.

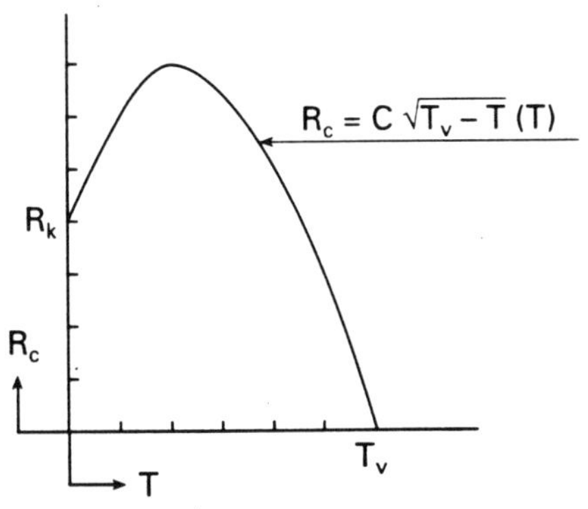

6.5 $R_c = f\,(T)$.

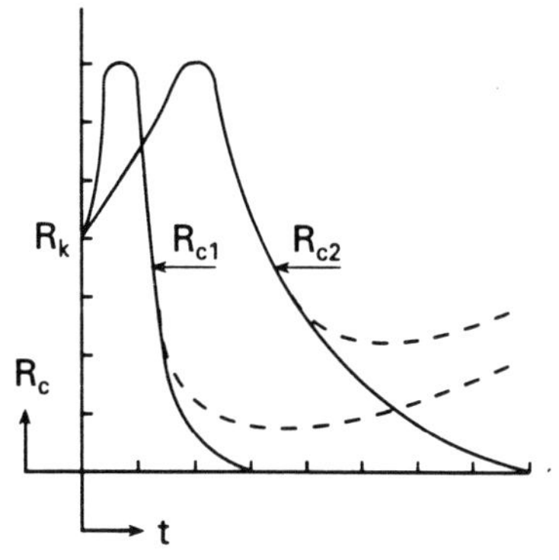

6.6 $R_c = f(t)$.

At a certain distance from the contact area, after the effect of R_C becomes negligible, material resistance ρ_e dominates and causes R to increase, see dashed line of Fig. 6.6.

6.6 Summary

Softening temperature, T_V, is often reached after the first millisecond, beyond which R_C has fulfilled its function, but even in this short period useful heat is generated in the weld which determines resultant weld properties.

This emphasises how important it is to achieve good process control during the first milliseconds. Use of DC equipment meets this requirement, see section 13.

7 ELECTRODES

7.1 Introduction

Resistance welding is a process in which the parts to be welded are squeezed together by the electrodes, which have to:

- Supply welding current;
- Press together the material to be welded;
- Cool the weld.

Electrodes have to have:

- Low electrical resistance;
- High thermal conductivity;
- Homogeneous material structure;
- No sticking or corrosion tendencies;
- High resistance to wear;
- Good hardness, thus sufficient mechanical resistance to deformation.

Some of these requirements are contradictory, e.g. a copper electrode has a low electrical resistance but the material easily deforms. A tungsten electrode, on the contrary, is deformation-proof but the sticking tendency is greater than with copper.

Compromises have to be made, and it is noticeable that beside the well known types available, there is hardly anything new. At the end of this section some special electrode materials will be examined.

7.2 Wear

A characteristic problem of resistance welding is electrode wear. As a result of wear the operational life of electrodes is restricted.

The most common wear is a result of deformation and mushrooming of the electrode tip. Current density decreases because of this and consequently the weld nugget diameter as well. The decrease in weld nugget diameter is sometimes indicated by a decline in welding voltage. When a preset minimum voltage has been reached, the welding machine switches to a higher welding current so that the current density remains constant. By doing so the operational life of the electrode is extended.

Welding of galvanized sheet gives alloying and sticking problems. This is compensated for by keeping the welding energy around the welding area as constant as possible. Often projection welding is used to overcome these problems, Fig. 3.3.

Welding operations in industry may encounter:

- Corrosion and sticking, for instance when welding tungsten wire to molybdenum;
- Appearance of grooves on the electrode surface when wire welding;
- Adhesion of oxide particles to the welding surface, for example, when welding TV tube shadow masks;

- Electrodes overheating caused by too high a current density;
- Contamination of electrodes by plastics vapours when welding foil capacitors.

7.3 Reducing electrode wear

The following points are important:

- Choose the right electrode material;
- Electrode material treatment such as drawing, hammering and hardening;
- Use of composite electrodes, for instance a hard wear resistant layer on a softer foundation. Hard particles such as oxides are absorbed in the layer without contributing unduly to wear. So far, not much experience has been gained using composite electrodes;
- Allowing for various effects such as Peltier and Thomson, which are polarity dependent, see section 12;
- Metallurgical aspects, inhomogenities, crystal structure or sintered composition, see section 7.6 about special electrode materials;
- The way in which welding energy is supplied greatly influences electrode wear, see DC equipment, section 13.

7.4 Electrode materials

To simplify choice of electrode materials, each welding machine manufacturer supplies a list of materials with their characteristics and applications. Often the standard RWMA classification (the Resistance Welder Manufacturers Association in the USA) is followed. This association divides the materials into two groups, see Table 7.I.

Others such as zirconium-copper alloy with characteristics which form a combination of class 1 and 2 did not find a place in the above classification of 1969. Nevertheless, they are supplied by many manufacturers.

Table 7.I Standard classification of electrode materials according to RWMA.

Group A Copper based electrodes.

Class	Composition, %	Application
1	Cadmium Cu 1/99	Aluminium, magnesium, mild steel
2	Chromium Cu 0.8/99.2	Various
3	Beryllium NiCoCu 0.5/1/1/97.5	Hard materials
4	Beryllium CoCu 1.8/0.3/97.9	Materials with sticking tendency, therefore large contact area, resistant to wear
5	Aluminium Cu 12/88	Materials needing a large tangent plane, good wear and sparking resistance

Group B Sintered electrode materials

Class	Composition, %	Application
10 11	Copper W 40/60 Copper W 32/68	Metals with high hardness, such as stainless steel
12	Copper W 22/78	As inserts for various applications
13	Pure tungsten(w)	For non-iron metals with high λ; smaller sticking tendency than for composite electrode materials. For welding Cu, CuZn and CuSn litz wire
14	Pure molybdenum(Mo)	As inserts for welding tombac, brass and silver. Moderate sticking tendency

More information on applications and forms in which the electrode materials are supplied is given in Table 7.II.

Table 7.II

RWMA class	Material	Philips norm	Composition, %	Melting temp, T_m, K	Density, tm^{-3}	Hardness, Vickers	Electrical resistivity, ρ_e, $\Omega m \times 10^{-8}$	Thermal conductivity, $Wm^{-1}K^{-1}$
1	HC Copper	NLN-R132	100 Cu	1356	8,9	100	1,8	385
1	HC Copper hard	NLN-R138	99,9 Cu	1356	8,9	95	2,0	322
2	Copper Chromium	NLN-R014	0,5 Cr rest Cu	1356	8,9	130	2,2	288
2	Copper Chromium Zirconium		0,4 Cr; 0,1 Zr rest Cu	1356	8,9	150	2,2	322
2	Zirconium Copper		0,2 Zr rest Cu	1356	8,9	150	2,0	340
3	Beryllium Cobalt Copper	NLN-R147	2,3 Co; 0,45 Be rest Cu	1356	8,9	220	3,2	175
12	Copper Tungsten	NLN-P046	25 Cu 75 W	1356	15,0	230	5,8	146
13	Tungsten ZG	LN-P096	99,9 W	3680	19,2		6,4	159
13	Tungsten ST	LE-P040	Th0 1,5 rest W	3680	18,2		5,6	128
13	Tungsten VT	LN-P073	Th0 3 rest W	3680	18,2			
13	Tungsten TT	LE-P087	Th0 0,75 rest W	3680	18,0		5,9	
13	Tungsten Y	D 2599	1 Y_2O_3 rest W	3680	18,3		5,9	
14	Molybdenum	LE-P001	99,9 Mo	2890	10,2	210	6,0	135

Special electrode materials

3	Nickel Beryllium Copper		0,5 Be; 2 Ni; 1,5 Zr rest Cu	1356	8,8	260	3,2	
	Silver Tungsten		40 Ag; 60 W	1233	14,5	170	3,0	260
2	Glidcop Al60	Weld-Equip	1,1 Al_2O_3 rest Cu	1356	8,8	200	2,2	300

7.5 Philips electrode materials

The Company uses electrode materials supplied by the warehouse No. 4500 of HIG-Machinefabrieken under the description Ph-norm. These materials are classified under 1-3 of Group A and 12 of Group B of RWMA. Tungsten and molybdenum are available from Ph-Maarheeze, Holland.

7.6 Special electrode materials

Sometimes one has to use electrode materials which are not available from one's own Company, for instance silver-tungsten 40/60. This material has good resistance to deformation, low resistance and moderate or negligible sticking tendency. At high current density the material meets the requirements less satisfactorily.

This and materials based on Mo are supplied by Bayerische Metallwerke GmbH, 7530 Pforzheim, Germany, Louis Renner, Schützenstraße 7, 806 München, Germany and Metallwerke Plansee AG, Reutte Tirol, Austria.

It is sometimes recommended to use monocrystalline tungsten, the characteristics of which are great hardness and resistance to wear together with low electrical resistance. Metal Research Imanco, Benelux Office, Postbox 578 in Tilburg, Holland supplies this material. The use of this material, however, has been restricted until now.

A material with greater potential is GlidCop A160 supplied by Glidden Metals Group, Cleveland, Ohio, USA represented in Holland by Weld-Equip Helmond. GlidCop material hammered from 0.375 inch to 2.5 mm diameter showed a three to fourfold life compared with the Mallory 328 electrodes used at present for welding 0.4 mm brackets and 0.2 mm grids of CrNiSteel. The material has a sintered structure of Cu grains, and Al grains which oxidise during sintering to Al_2O_3. As a result of sintering the hardness remains stable at a higher temperature than for material such as CrCu. Moreover, electrode hardness remains the same as its original value after use.

With conventional electrodes based on Cu, tip hardness decreases after use. This gives GlidCop material favourable characteristics that contrast well with traditional materials.

In 1977 the quality of some Japanese electrode material of composition 0.5% Be, 2% Ni, 1.5% Zr, Cu to 100% was examined. Metallurgical research indicated that no more could be expected from this material than Mallory 100 or CoCu.

7.7 Electrode shape

When welding two sheets of different thicknesses the electrode combination of Fig. 7.1 is chosen so that, when current flows, the weld nugget is biased towards the thinner sheet. Alternatively, the electrode in contact with the thinner sheet is made from material which holds the heat more than the other electrode, see Fig. 7.2.

With choice of electrode dimensions, composition and shape, heat distribution is controlled as required.

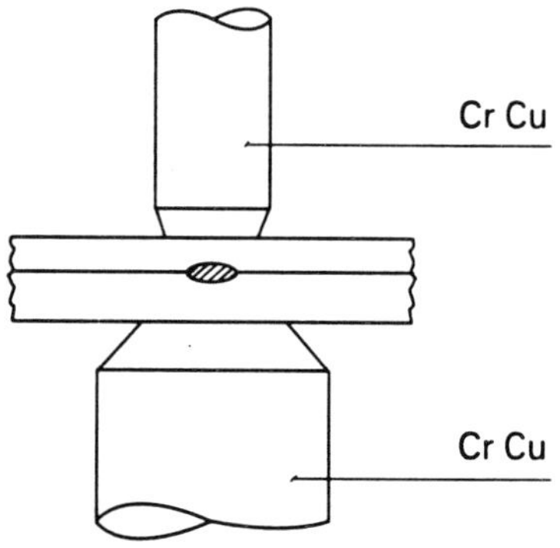

7.1 Improved heat distribution via electrode dimensions.

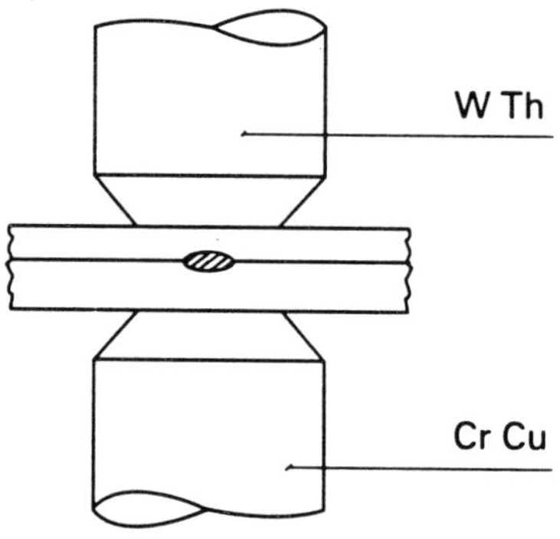

7.2 Improved heat distribution via electrode composition.

The most commonly used electrode tips are shown in Fig. 7.3.

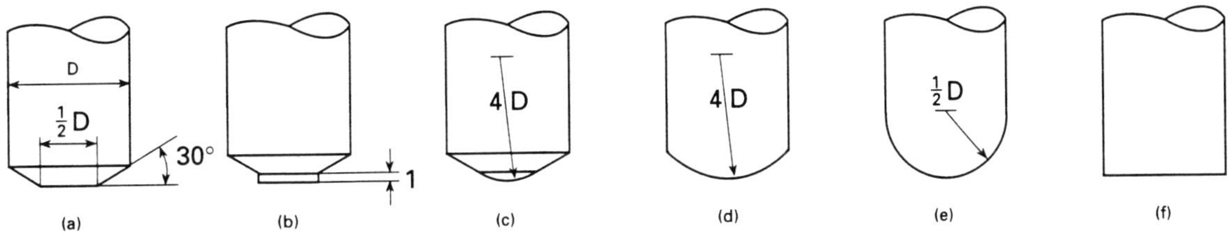

7.3 Electrode tip shapes: a) Truncated cone; b) Truncated cone with pimple; c) Domed truncated cone; d) Domed; e) Convex; f) Flat.

It is not easy to say which shape is the best. Factors such as material to be welded, whether Al, NiFe or Cu, surface condition and hardness and roughness of the material to be welded all influence the chosen electrode shape. These effects have to be analysed experimentally, and the results are self-evident. Oxidised steel sheet is welded more easily with electrode (c) Fig. 7.3 than (f).

On the other hand when welding Al sheet, according to work done by The Welding Institute,[2] the best results were obtained with the electrode of Fig. 7.3a). During an experiment regarding welding 0.2 mm sheet to 0.4 mm sheet, both of CrNiFe, the dome of the electrode tip, Fig. 7.3d), greatly influenced operational electrode life.

In short, the best electrode shape has to be determined by experiments and is dependent on welding conditions. A pointed electrode concentrates welding energy better than a flat electrode, but the heat load is greater. For material with a tin layer or oxide film it is particularly recommended to concentrate energy using projections on the material to be welded, Fig. 7.4. The electrode tip remains flat, which results in minimum heat load in the electrode.

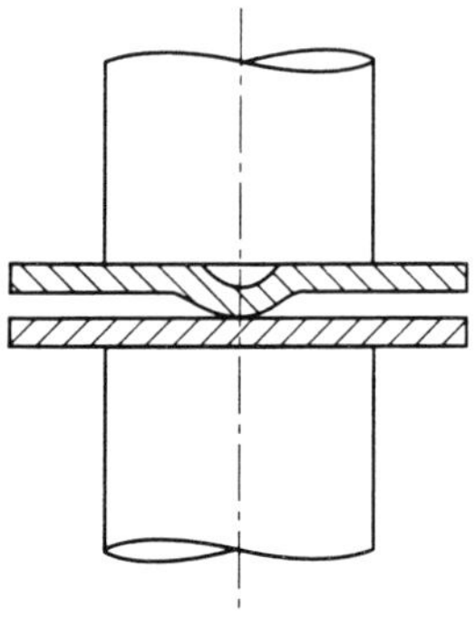

7.4 Projection weld.

8 MEASURING

8.1 Introduction

The expression 'measuring is knowing' conceals difficulties faced when measured results have to be assessed to obtain the correct value. Recording equipment for the event to be measured normally produces a voltage pulse which is integrated, amplified and stored in memory before the result is displayed.

To get accurate representation of the event requires that its capture and processing, from signal to visual display, must be correctly performed.

What does the signal mean? If the welding pulse is, for instance, processed as if it were a complete sinusoid, with i_o as peak value, Fig. 8.1, then i_o is a known quantity for the energy
$\frac{W}{R} = \left(\frac{i_o^2}{2}\right)t$ which increments per unit of resistance.

It becomes more difficult when the weld is made with a chopped sine wave, Fig. 8.2. How is i_o translated with respect to the energy supplied? Also, that the welding machine is earthed sometimes causes difficulties in measuring current and voltage which need to be identified and allowed for. Finally, problems may originate at the factory because basic information is not made available through lack of measurement experience.

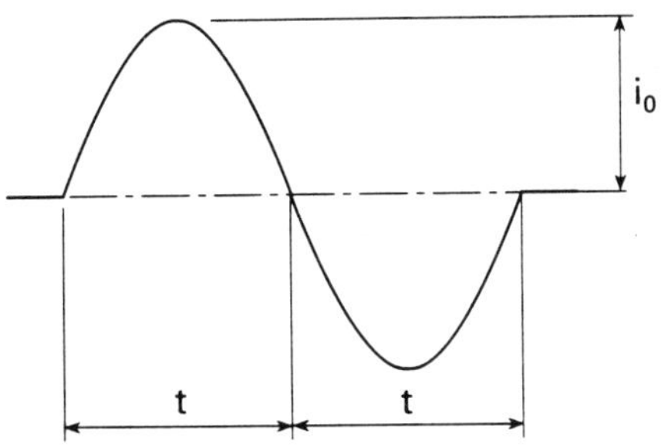

8.1 Complete sinusoid.

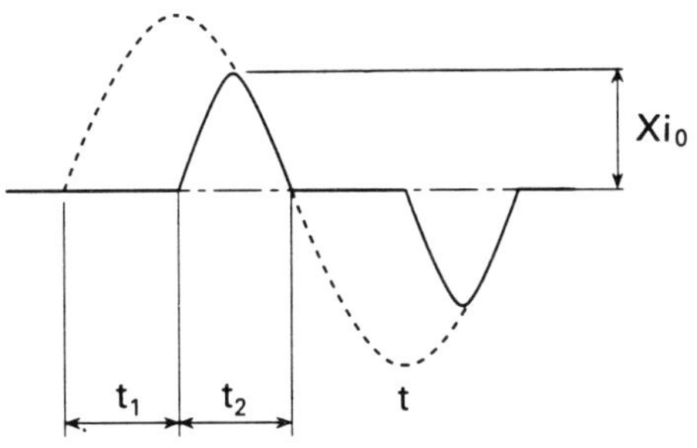

8.2 Chopped sinusoid.

The author therefore found it appropriate to devote one section to measurement techniques, in particular to measurement of:

- Current;
- Voltage;
- Resistance and capacitance;
- Movement;
- Welding force.

8.2 Current

1 Equipment

Welding current pulses are usually measured with a meter driven from a toroidal induction coil which is placed around the current conductor. The induced voltage pulse is integrated and displayed in analog or digital form.

Suppliers of suitable equipment include Weld-Equip BV, Helmond, Holland and Messer Griesheim Nederland BV, Amsterdam, Holland.

As long as current pulses are of a complete sinusoidal waveform, this measuring method is reliable. As already mentioned in the introduction, measurement becomes more critical as the pulses to be measured become progressively chopped.

Imagine that $t_1 = t_2 = \frac{1}{2}t$ of Fig. 8.2 becomes

$$Xi_o = i_o \sin 3/4\pi = 0.71\ i_o$$

For a complete sinewave the energy during a half period is

$$\frac{W_1}{R} = \frac{i_o^2}{2}(t)\ J/\Omega$$

For a chopped sinewave the energy during a half period is

$$\frac{W_2}{R} = \frac{(0.71 i_o)^2}{2}\ (\tfrac{1}{2}t) = \frac{0.25 i_o^2}{2}\ (t)\ J/\Omega$$

Thus $W_2/W_1 = 0.25$.

The current pulse peak value falls from i_o to 0.71 i_o and the corresponding welding energy from W_1 to 0.25 W_1.

Measurements show that at not too great a phase shift, the measuring equipment integrates the current pulse fairly well over the time during which measurement is made. When measuring a welding pulse consisting of more than one cycle, that part with the maximum amplitude is assumed, as a rule, to relate to the complete pulse.

Peak values of welding pulses from 1000 Hz and DC capacitor discharge equipment are of main interest; upslope is not usually measured.

In conclusion, when welding with 50 Hz alternating current it is sensible to use sinewave pulses as far as possible.

2 Current shunt and oscilloscope

A reliable method displays the measured signal on an oscilloscope with memory. The voltage signal corresponding to the current to be measured is derived from a low inductance shunt in the secondary circuit, Fig. 8.3.

The low inductance shunt is used for frequencies of ≤ 50 Hz, and also for DC. For frequencies of ≥ 1000 Hz an induced voltage component is present and must be allowed for when analysing the signal obtained. Furthermore, whether and where the welding machine is earthed are also important.

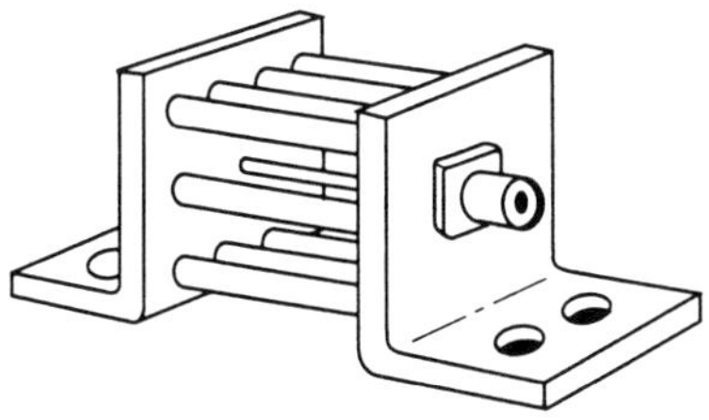

8.3 Low inductance resistive current shunt.

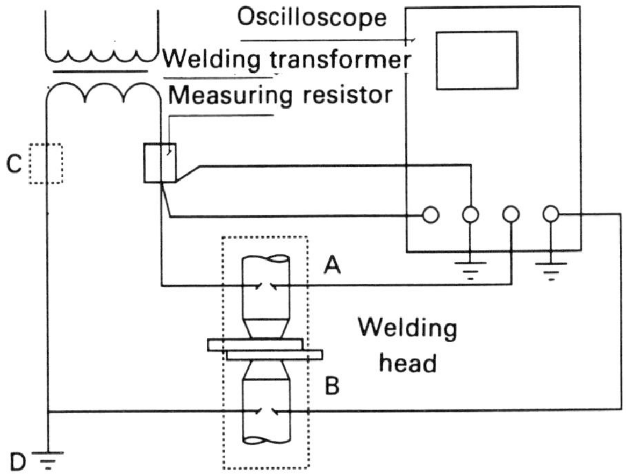

8.4 Measuring circuitry.

When, for instance, current is measured via the resistive shunt of Fig. 8.4, measured current is higher than the welding current, as part of this leaks to earth. It is therefore better to place the resistance at C.

The same applies to measurement of welding voltage. For the set-up in Fig. 8.4, the voltage difference A-D is measured instead of A-B. With these considerations in mind, use of a differential input oscilloscope without earthed input connections is recommended.

Measurement of low voltages may cause difficulties, however, as a result of pick-up and induced interference. Measuring with an oscilloscope has the advantage that behaviour is visually displayed.

3 DC power supply

When using a DC welding unit with stabilized current adjustment, measuring the current pulse is superfluous as pulse conditions are read from the welding unit controls. In this case, welding voltage is a good reference quantity for what is happening in the weld.

8.3 Voltage

Measuring welding voltage is a simple operation which gives good operating information. Sometimes welding voltage parameters are used to control weld nugget appearance, if present, see section 14.3e). Other indications are that welding force is controllable using information derived from welding voltage, variations in current pulse and welding time. Measuring the welding voltage has interesting possibilities in combination with other parameters. For instance, when using constant current pulsing, especially a DC pulse, the weld can be monitored by measuring voltage, see Fig. 8.5 and 8.6.

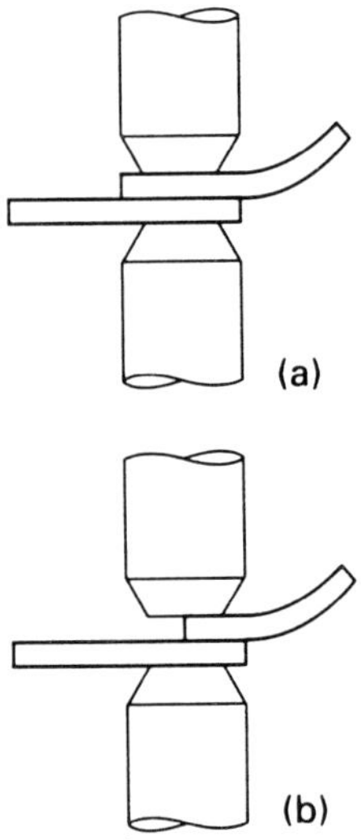

8.5 a) Good weld; b) Bad weld.

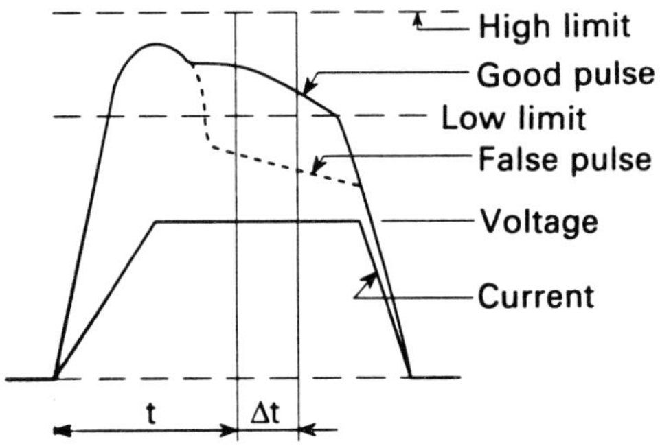

8.6 Measurement period. Good pulse of Fig. 8.5a Bad pulse of Fig. 8.5b.

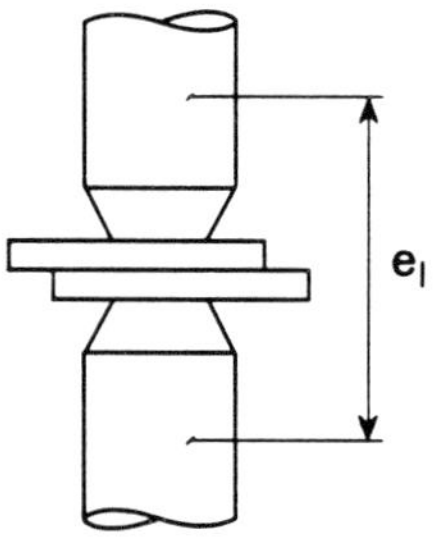

8.7 Voltage measurement across the electrodes.

Bad welds are eliminated by measuring voltage during t, see Fig. 8.6. It is difficult to measure welding voltage across the weld itself, usually measuring across the two electrodes is acceptable, see Fig. 8.7.

8.4 Resistance and power

Current and voltage are better measured separately for good results, but measuring dynamic resistance R = e/i and power P = ei have, until now, presented difficulties as both are zero at the beginning and end of a pulse.

There are, in fact, dynamic resistance meters but they are made for welding times of many cycles at 50 Hz and peak values only of these are measured. In microwelding this is not of interest because it is the behaviour of the resistance and performance at the start of the pulse which matter more.

Use of a microcomputer opens up new ways of controlling R and P during welding.

For measuring static resistance, such as contact resistance at room temperature, R_k, and material resistance ρ_e, a DC microresistance meter is used, for instance, the meter supplied by Tekelec Airtronic, Cité des Bruyères in France, and their representative Tekelec Airtronic NV in Amsterdam.

8.5 Movement

Measuring displacement of the movable electrode during welding, in combination with the current pulse, contributes a great deal to good analysis of the welding process. This applies especially when there is a reasonable movement, for instance when butt welding wire or sheet.

For movement of a flat cut wire as shown in Fig. 8.8, expansion of the material is registered at the beginning of the welding pulse and at the moment that the wire end softens. From this movement an impression of the welding dynamics is gained, see section 9.

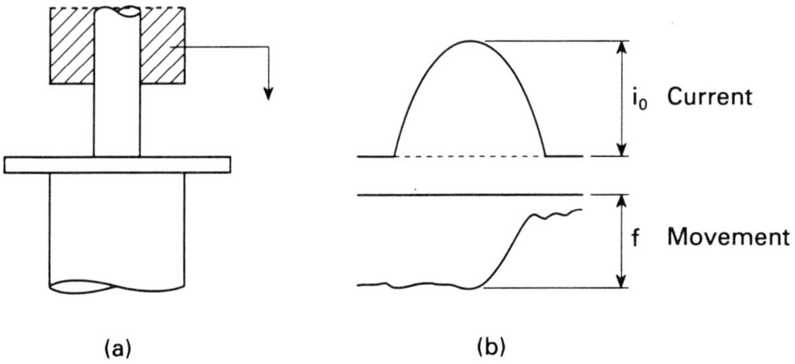

8.8 Measuring displacement.

When the wire is pointed, Fig. 8.9, movement takes place from the beginning of the welding pulse.

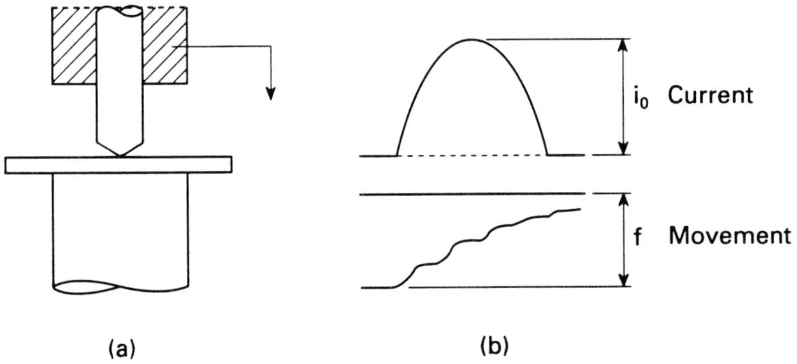

8.9 Measuring movement.

For projection welding this measuring method is particularly informative.

For measuring movement the following may be used:

- Fotonic Sensor of MIT Instruments New York, USA, representative Lindeteves-Jacoberg Nederland BV, Amsterdam, Holland;
- The eddy current movement meter of Kaman Measuring Systems USA represented by Airparts Alphen ad Rijn, Holland.

These are contactless measuring systems.

8.6 Welding force

Measurement of welding force gives the researcher a good idea of the dynamic behaviour of the welding head, see also section 9. On rapid heating, rapid material expansion produces a correspondingly fast rise in welding force, sometimes at resonance. Depending on the damping and natural oscillation frequency of the system, these variations can exist for some time.

If the material to be welded becomes soft at the weld spot or fusion occurs, then the movable electrode is displaced, resulting in a decrease in force. When this decrease in force corresponds to the nominal welding force, current interruption takes place, Fig. 8.10.

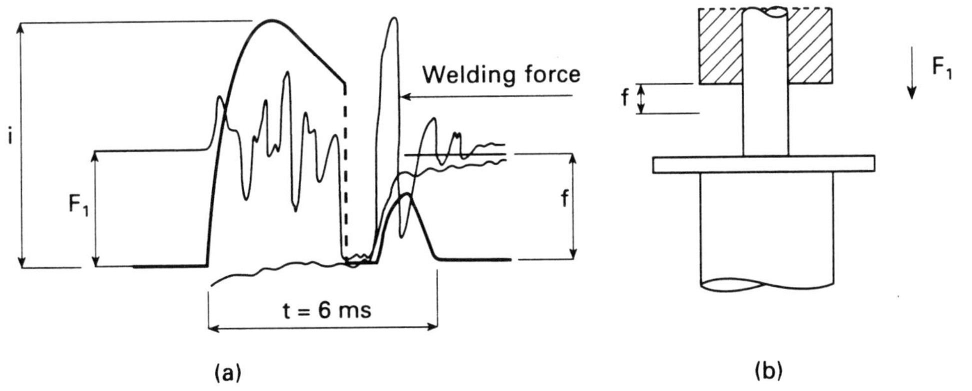

8.10 a) Interruption of welding current as a result of $\Delta F = F_1$ causes sparking, b) Weld set-up.

Splash is caused by molten metal being ejected from the fusion spot. For measuring welding force a ring of piezoelectrical material is used which generates a voltage when subjected to changing pressure. This voltage is amplified and saved in an oscilloscope memory, Fig. 8.11.

A supplier of suitable equipment is Kistler Instruments AG, Winterthur, Switzerland, represented by Geveke Fysica BV, Amsterdam, Holland.

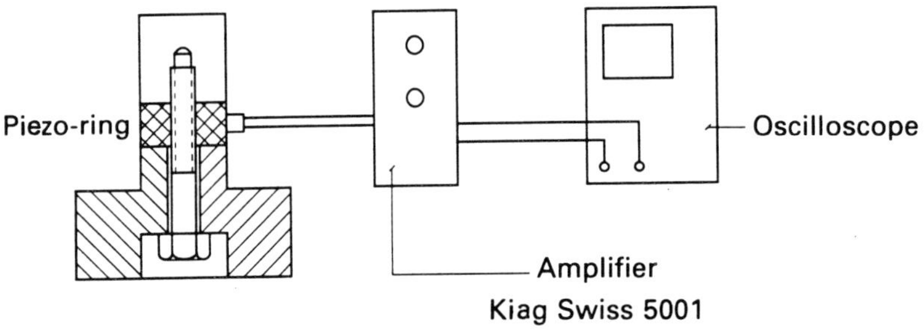

8.11 Force measuring set-up.

9 WELDING DYNAMICS

9.1 Introduction

Welding force, F_l, considerably influences the welding process. Two functions of F_l must be distinguished:

- At the beginning of the welding pulse, static F_l determines the contact resistance R_c, see Fig. 9.1.

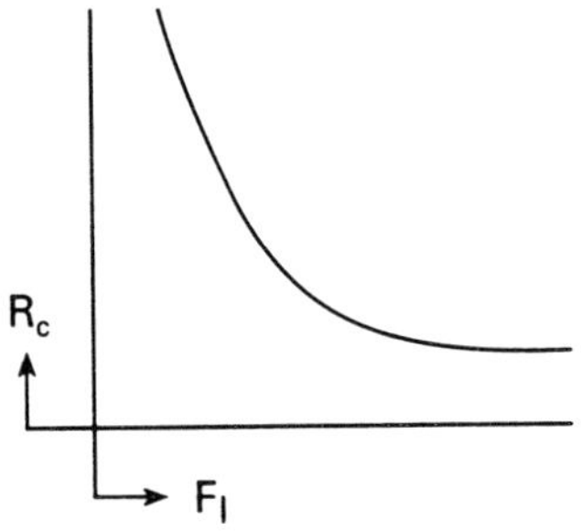

9.1 R_c as a function of static F_l.

As F_l is increased, R_c decreases proportionally. Above a limiting value of F_l, R_c does not change further. Thus the beginning of F_l determines the heat concentration in the weld, see also section 6.

- As the welding pulse progresses R_c decreases quickly as the material becomes plastic.

It is important to keep the weld components and electrodes in close contact, and sufficiently accelerate the movable parts of the weld head. This is effected by dynamic F_1.

In this section the dynamics of the welding process are analysed with the aid of a dynamic model and explained using some examples.

9.2 Dynamic F_1

A starting point for easily measurable welding parameters is the weld set-up as shown in Fig. 9.2.

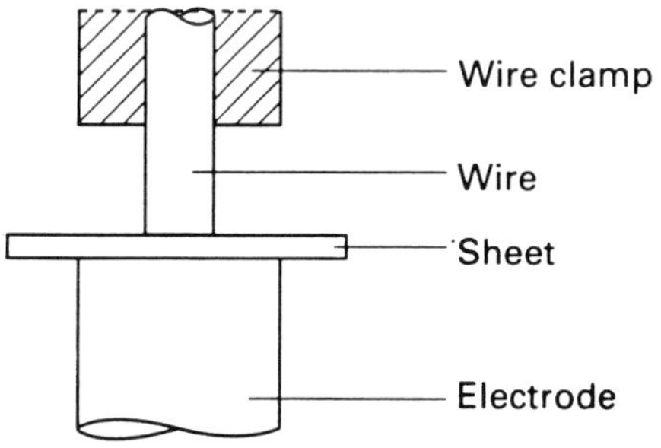

9.2 Weld set-up.

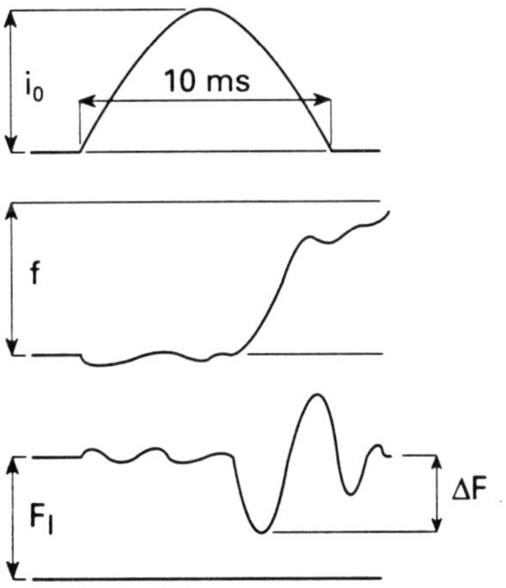

9.3 Measured weld quantities.

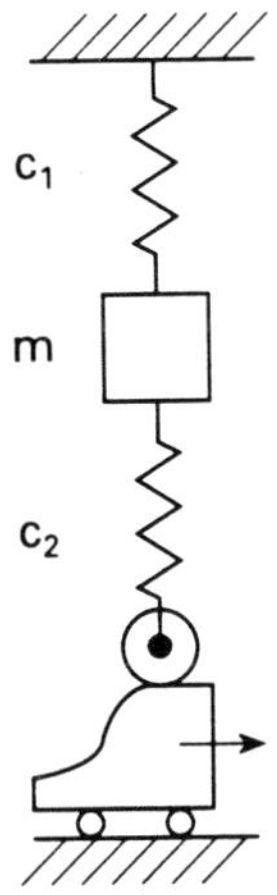

9.4 Dynamic model.

Figure 9.3 shows: the current pulse, i, (half cycle), movement f and welding force, F.

Figure 9.4 represents a dynamic model of the set-up of Fig. 9.2, where c_1 = stiffness of the compression spring, m = mass of the movable parts and c_2 = stiffness of the material to be welded.
At the moment that current starts to flow, attention is paid to two aspects:

- The compression spring;
- The mass-spring system of the weld.

9.3 Compression spring

What requirements does the spring have to meet? Figure 9.5 helps to answer this question.

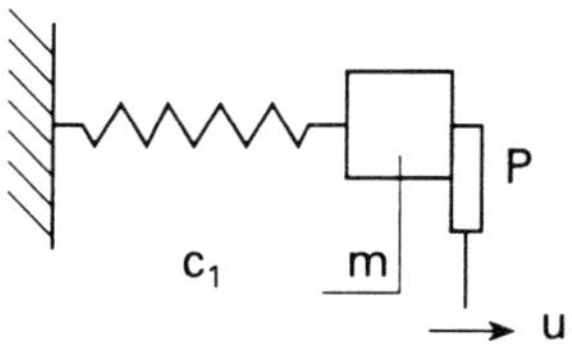

9.5 Mass-spring model.

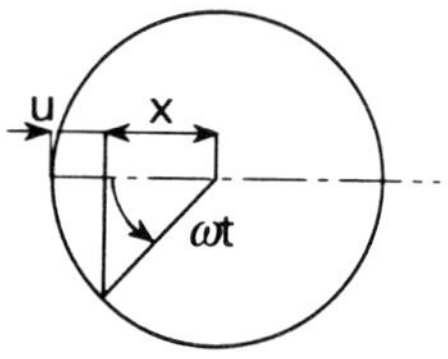

9.6 Vibration.

When the material to be welded softens, it is like releasing pawl P restraining m, so that the compression spring is freed. Mass m starts vibrating at its natural frequency according to

$$m\ddot{x} + c_1 x = 0$$

Only the beginning of vibration is taken into account as damping immediately follows, caused by the material to be welded. The vibration is expressed by

$$x = x_o \cos \omega t$$
$$u = x_o - x = x_o (1 - \cos \omega t)$$
$$\dot{u} = x_o \omega \sin \omega t$$
$$\ddot{u} = x_o \omega^2 \cos \omega t, \text{ see Fig. 9.6.}$$

The lifting off condition for the movable electrode is

$$\ddot{u} = x_o \omega^2 = x_o \frac{c_1}{m} = \frac{F_1}{m}$$

It is necessary that mass m is low to achieve sufficient acceleration u.

The force $F_1 = x_o c_1$ must be big enough and as constant as possible during movement. This means that the stiffness is low. In short a low mass m, minimum friction, and low spring stiffness are needed.

9.4 Mass-spring system

What is the dynamic system in resistance welding? Taking into account what has been said already, the model of Fig. 9.4-9.7 can be simplified.

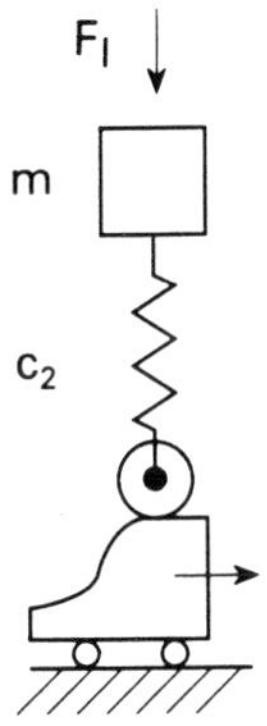

9.7 Simplified dynamic model.

For a weld set-up like Fig. 9.2, the material will fuse or soften after time t, resulting in large oscillations in welding force F_l, see Fig. 9.3. When the change in force, ΔF, becomes too big or equal to F_l, then current interruption and sparking occur.

Therefore it is important to know something more about the welding dynamics. The movement of the cam roller, Fig. 9.4 is the mechanical equivalent of a surge in weld spot growth, in velocity and acceleration. Derivation of the equation is given in [9.1]. Calculation of the decrease in force according to these three methods will be shown on pages 46,47 and 49. There follows a simplified method for calculation of the decrease in force, ΔF_l, which best represents the surge in acceleration:

$$V_{max} = \frac{h}{t_h}$$

$$a = \frac{V_{max}}{t_a} = \frac{h}{t_h \; t_a}$$

$$x_{max} = 2a = \frac{2h}{t_h \; t_a}$$

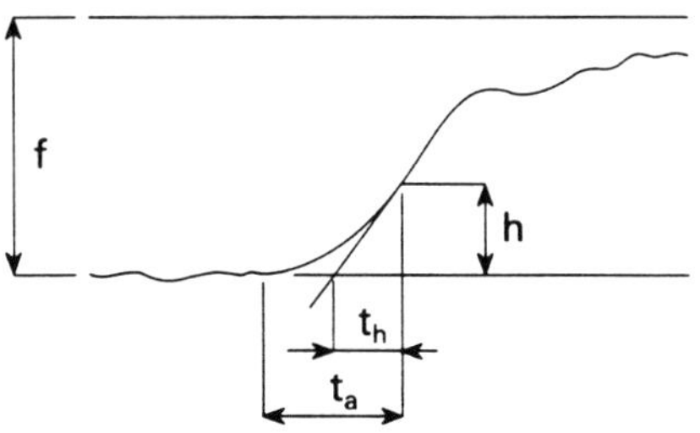

9.8 Movement of electrode.

Example:

$h = 80.5\ \mu m$

$t_h = 0.7$ ms

$t_a = 1.12$ ms

$a = 102.5\ m/s^2$

$\ddot{x} = 205\ m/s^2$

For m = 100 g, $\Delta F_1 = 0.1 \times 205 = 20.5$ N.
The change in welding force is calculated according to the amount of movement.

9.5 Other effects

1 Current force
For a straight cut wire, Fig. 9.1, there may be an instant during which electrode movement is too slow to match electrical energy input, which continually increases, resulting in current interruption and sparking.

2 Mass
Too great a mass may lead to current interruption but the effect is not linear, $\Delta F_1 = f(\sqrt{c})$, see equation [3], earlier. It is always recommended, especially for high speed machines, to keep the movable mass as low as possible.

3 Upslope gradient of welding pulse
Experiments show that a steep upslope of the welding pulse causes large force variations. These large changes in welding force often dominate other effects and no longer fit in the above dynamic system models.

4 The welding force itself does not have any influence on the variations but F_1 has to be adjusted so that it is always bigger than the negative ΔF_1.

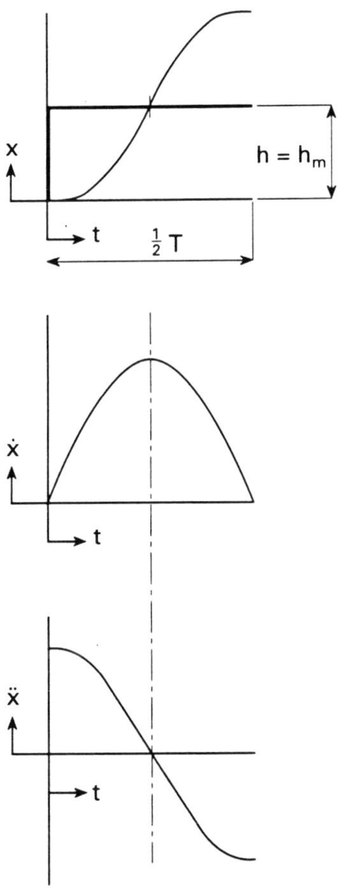

9.9 Increase in movement.

The constrained increase in movement is as follows:

$$h\ (t) = h_m$$

The resulting movement is then:

$$x = h_m\ (1-\cos\omega_e t)$$

$$\dot{x} = \omega_e h_m\ \sin\omega_e t$$

$$\ddot{x} = \omega_e^2\ h_m\ \cos\omega_e t$$

$$\omega_e = \frac{2\pi}{T}$$

The maximum value of x stands for:

$$t = 0 \longrightarrow \ddot{x}_{max} = \omega_e^2\ h_m \quad \ldots [4]$$

$$\Delta F_1 = \ddot{x}_{max}\ m = \omega_e^2 h_m m \quad \ldots [5]$$

Table of two practical examples (see earlier)

Quantity	Case 1	Case 2	Units
Mass, m	$17x10^{-3}$	$108x10^{-3}$	Kg
Weld pulse, i	873	876	A
Weld time, t	10	10	ms
Weld force, F_1	15	12	N
2 h_m	161	232	μm
½T	2.24	4.1	ms
ω_e =2π/T	$1.4x10^{-3}$	$0.77x10^{-3}$	
$\ddot{x}$ max	158	69	m/s²
ΔF_1	2.7	7.5	N

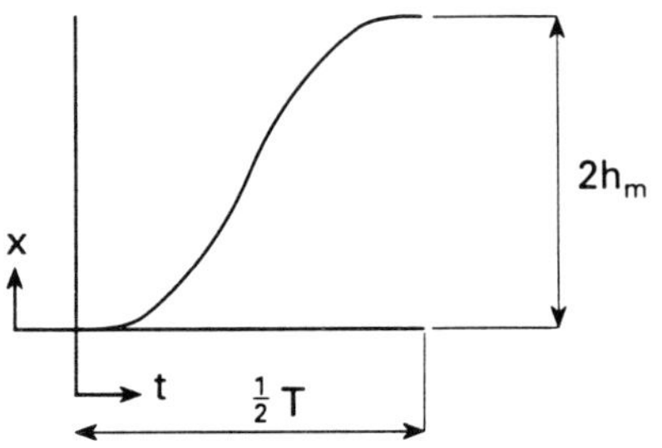

9.10 The resulting movement.

Remark: The values of F_1 are slightly lower than the starting point of the leap in velocity.

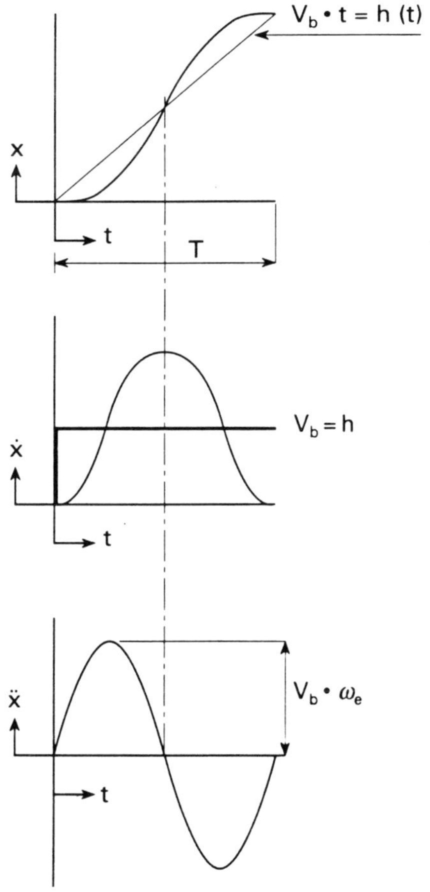

9.11 Increase in velocity.

For the increase in velocity the constrained movement is:

$$\dot{h} = V_b$$

$$h\ (t) = V_b t$$

The resulting movement is:

$$x = V_b\ t - \frac{V_b}{\omega_e}\ \sin\omega_e t$$

$$\dot{x} = V_b\ \ (1-\cos\omega_e t)$$

$$\ddot{x} = V_b\ \omega_e\ \sin\omega_e t$$

Here $\omega_e = \frac{2\pi}{T}$

The maximum value of x for $t = \frac{1}{4}T$ is $\ddot{x}_{max} = V_b\ \omega_e$

. . . [1]

$$\Delta F_1 = \ddot{x}_{max}\ m = V_b\ \omega_e\ m$$

. . . [2]

$$\omega_e = \sqrt{\frac{c}{m}} \qquad \boxed{\Delta F = V_b\ \sqrt{cm}}$$

. . . [3]

c here is the stiffness of the material to be welded.

Equation [3] shows that the force, decrease in ΔF_1, is proportional to $\sqrt{m}$ and not m.

Two more practical examples

Quantity	Case 1	Case 2	units
Mass, m	$17x10^{-3}$	$108x10^{-3}$	kg
Weld pulse, i	873	876	A
Weld time, t	10	10	ms
Weld force, F_1	15	12	N
h	161	232	μm
T	2.24	4.1	ms
V_b = h/T	$72x10^{-3}$	$56.5x10^{-3}$	m/s
ω_e = $2\pi/T$	$2.8x10^{-3}$	$1.53x10^{-3}$	
$\ddot{x}_{max}$	201	86.5	m/s^2
$\Delta F_1 = \ddot{x}_{max}\ m$	3.42	9.35	N

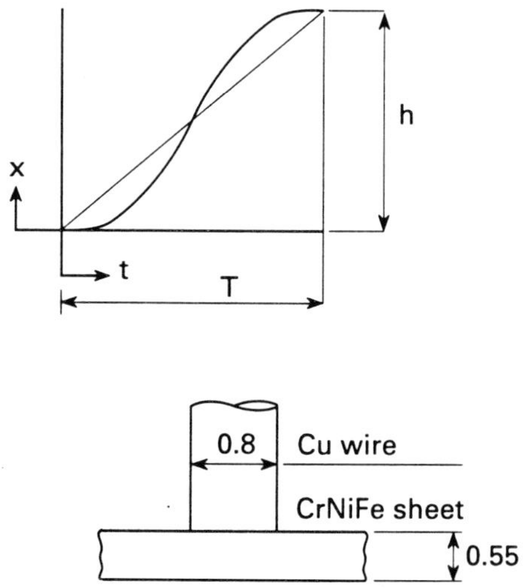

9.12 Welding conditions.

The change in force is:

$\Delta F_{12}/\Delta F_{11}$ = 2.7 which approximates $\sqrt{m_2/m_1}$ = 2.5

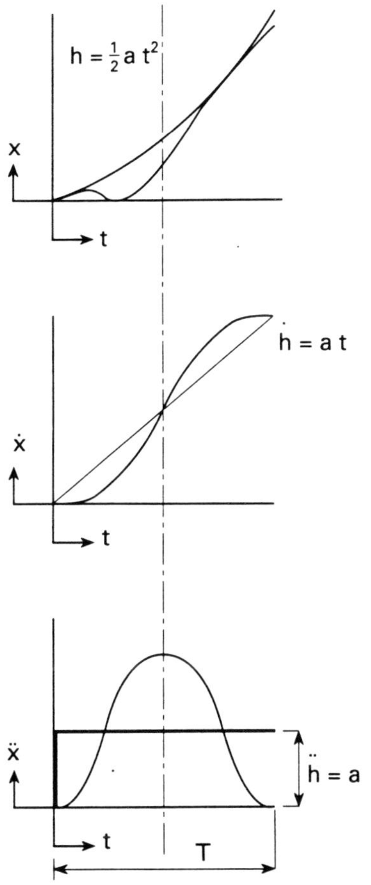

9.13 Velocity increase.

For the increase in velocity the constrained movement is:

$h = \frac{1}{2} at^2$

$\dot{h} = at$

$\ddot{h} = a$

The resulting movement is:

$x = \frac{a}{\omega_e^2} \cos \omega_e t - \frac{a}{\omega_e^2} + \frac{1}{2}at^2$

$\dot{x} = -\frac{a}{\omega_e} \sin \omega_e t + at$

$\ddot{x} = a\ (1-\cos \omega_e t)$

$\omega_e = \frac{2\pi}{T}$

The maximum value of $\ddot{x}$ applies to

$t = \frac{1}{2} T = \frac{\pi}{\omega_e} \qquad \ddot{x}_{max} = 2\ a$ . . . [6]

$a = \frac{2h}{T^2}$ . . . [7]

$\Delta F_1 = 2am = \frac{2h}{T^2}\ 2m$ . . . [8]

Note: when using the increase in velocity, only the initial movement can be analysed.

Further practical examples

Quantity	Case 1	Case 2	Units
Mass, m	$17\text{x}10^{-3}$	$108\text{x}10^{-3}$	kg
Weld pulse, i	873	876	A
Weld time, t	10	10	ms
Weld force, F_l	15	12	N
h	80.5	116	μm
T	1.12	2.05	ms
$a=2h/T^2$	128.4	55.2	m/s^2
$\ddot{x}_{max} = 2a$	257	110	m/s^2
$\Delta F_l = \ddot{x}_{max}\ m$	4.37	11.9	N

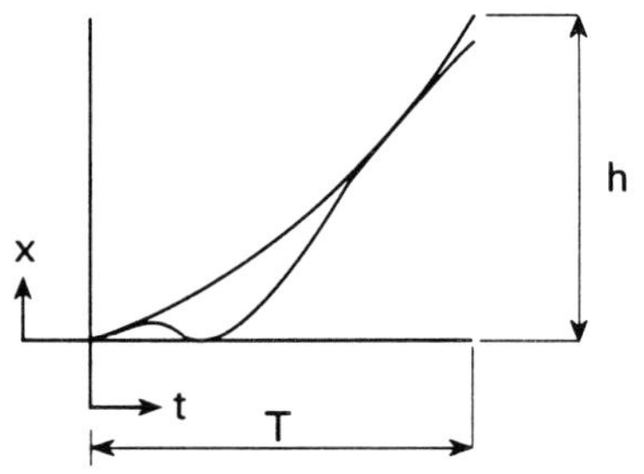

9.14 Initial movement.

Note: values of F_l are in this case higher than for the increase in velocity.

10 THERMAL MANAGEMENT

10.1 Introduction

Although normally there are few problems with resistance welding, the process is rather complicated. There are many factors which influence it and some are difficult to identify experimentally.

Beside the usually straightforward measurement of current, voltage, movement and welding force, it is temperature in particular from which useful information is obtained about what is happening in the weld. To measure temperature in the past an IR meter from Barnes Engineering Company, Conn, USA was purchased. With this meter it was possible to follow fast temperature changes. There are difficulties however, in that the weld spot is not easily accessible, the heating point is moving and emission of IR from the surface may change through oxide formation. This method was inadequate for our task.

There was also the possibility of calculating temperature by the PHILPAC computer program. This program is used for designing and analysing electronic circuits.

After the physical aspects of resistance welding are converted into an electrical analog, the program is ready for use.

It is possible to calculate, depending on factors entered, the temperature at any time and location. The results need critical assessment, as the computer sees only numbers, and processes only what has been entered.

Before using the computer a good knowledge of the physical operation, supported by measurements, is indispensable. Then analysis of conditions previously impossible may be performed. An example is measurement of contact resistance and its contribution to operation of the process.

In any case the temperature calculation permits evaluation of conditions otherwise impossible without this method.

In this section some of the most important results are dealt with.

The analogy is between Kirchoff's first law, which states that the algebraic sum of currents flowing at any junction in a network is zero, and the first law of thermodynamics, see section 10.2.

10.2 Analogy of thermal with electrical processes

The first law of thermodynamics states that total energy of a thermodynamic system remains constant although it may be transformed from one form to another. Here, heat developed = heat stored + heat dissipated, per unit volume.

$$j^2 \rho_e = \rho_c \frac{\delta T}{\delta t} - \lambda \frac{\delta^2 T}{\delta x^2}$$

The minus sign indicates heat loss.

Rearranging: $$\rho_c \frac{\delta T}{\delta t} = \lambda \frac{\delta^2 T}{\delta x^2} + j^2 \rho_e$$

Where

c	=	specific heat capacity,	J/m^3K
T	=	temperature,	K
t	=	time,	s (mostly ms)
λ	=	thermal coefficient of conductivity,	W/m/k
x	=	distance,	m
j	=	current density = i^2/unit area,	A/m^2
ρ_e	=	resistivity,	Ωm

In this formula details obtained by experiment and also details of R_c can be incorporated, see section 6.

The thermal circuit is divided into a number of sections and for each of them the first law[1] is applicable.

The PHILPAC program operates, however, with electrical quantities. The expressions used in the above mentioned law are therefore replaced by their electrical counterparts. A schematic representation of an electrical junction is shown in Fig. 10.1.

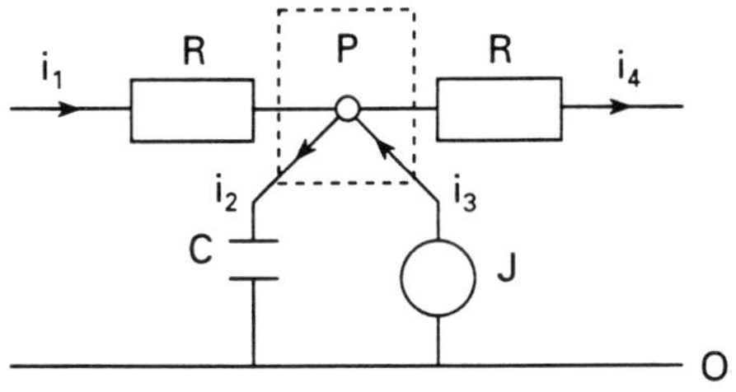

10.1 Schematic diagram of electrical junction, P.

According to Kirchoff's first law $\quad i_p = 0$

$$i_1 + i_3 = i_2 + i_4$$

$$i_1 - i_4 = i_2 - i_3 \qquad \ldots [2]$$

$$\Delta i_{1.4} = \Delta \frac{\Delta U}{R} = \Delta \frac{\Delta T}{R_{th}} = \Delta \frac{\Delta T}{\rho_e \frac{x}{A}} = \Delta \frac{\Delta T}{\frac{1}{\lambda} \frac{x}{A}}$$

$$\Delta i_{1.4} = \lambda \frac{\delta^2 T}{\delta x^2} \Delta V \qquad \ldots [3]$$

$$i_2 = \frac{\Delta q}{\Delta t} = C \frac{\Delta U}{\Delta t}$$

$$C_{th} = \rho c \Delta V \quad i_2 = \rho c \frac{\delta T}{\delta t} \Delta V \qquad \ldots [4]$$

$$i_3 = \frac{\Delta Q}{\Delta t} = \frac{\Delta i^2 R t}{\Delta t} = i^2 R = j^2 \rho \Delta V \qquad \ldots [5]$$

[3], [4] and [5] filled in [2] gives:

$$\Delta \frac{\delta^2 T}{\delta x^2} \Delta V = \rho c \frac{\delta T}{\delta t} \Delta V - j^2 \rho \Delta V$$

or also:

$$\rho c \frac{\delta T}{\delta t} = \lambda \frac{\delta^2 T}{\delta x^2} + j^2 \rho \qquad \ldots [1]$$

This derivation is in 10.1.

Thus with PHILPAC it is possible to calculate the analogous thermal circuit. It has to be mentioned, however, that until now a current source with i = constant has been assumed. It may sometimes be desirable to enter current as a dependent variable. In that case other computer programs will suit better.

10.3 Some applications

1 Contact resistance R_C

The R_C calculated in section 6 has a dominant influence on formation of the resistance weld, but what is its effect? To quantify this the weld set-up of Fig. 10.2 with two sheets to be welded each of 1 mm thickness is taken as a starting point.Although R_C exists momentarily only some ms, the temperature V10 of Fig. 10.2 increases considerably during this short time but the corresponding temperature without a significant R_C, e.g. R_C = 1$\mu\Omega$, is much lower, see Fig. 10.3.

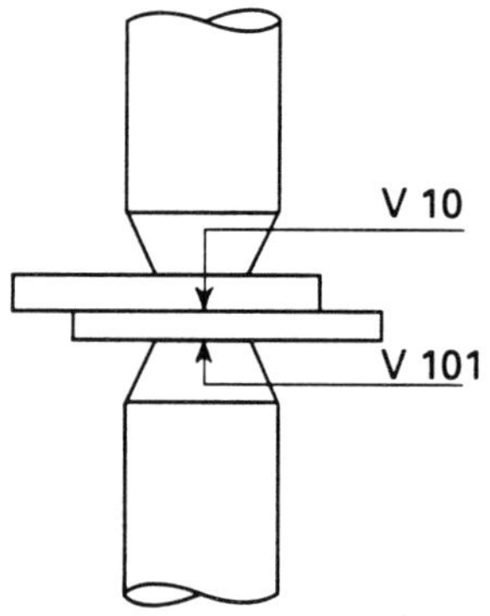

10.2 Weld set-up.

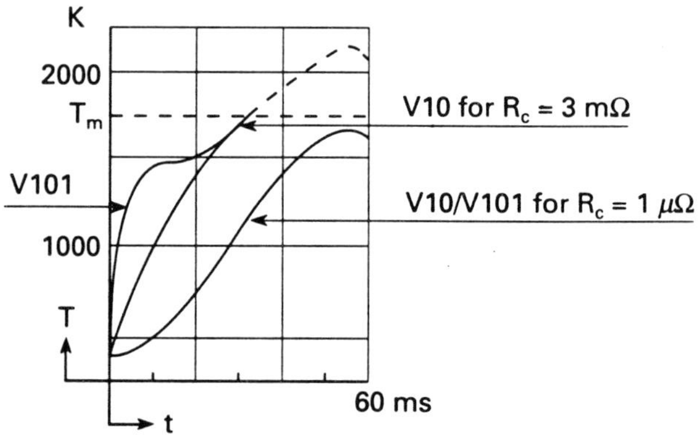

10.3 Temperature rise.

This shows that weld temperature development is predominantly caused by interaction between R_C and ρ_e.

For a material such as copper the influence of R_C is much smaller because of its lower resistivity, ρ_e.

2 Dimensions

The welding programme was first set up for welding steel of 0.5 mm thickness using electrodes of 5 mm tip diameter.By increasing effective current, i_{eff}, from 3000-9000 A the maximum temperature in the weld increased only from 1290-1360 K. This clearly shows that an electrode tip diameter of 5 mm, for sheet thickness of s = 0.5 mm was much too big. The electrodes cooled the weld too much and the energy was difficult to concentrate. This sometimes occurs when welding lids or tops on a flat surface. A large energy input is needed for a short period.

3 Weld heating

Another significant effect occurred when, with an increase in effective welding current, i_{eff}, heat production in the material interface actually decreased, see Fig. 10.4 and 10.5.

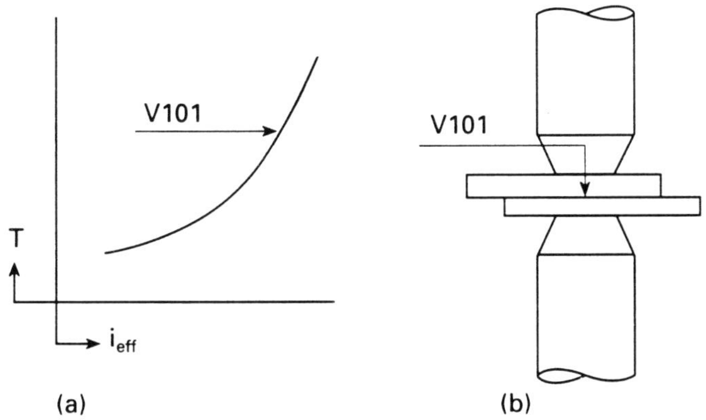

10.4 Temperature V101 increases with increasing i_{eff}.

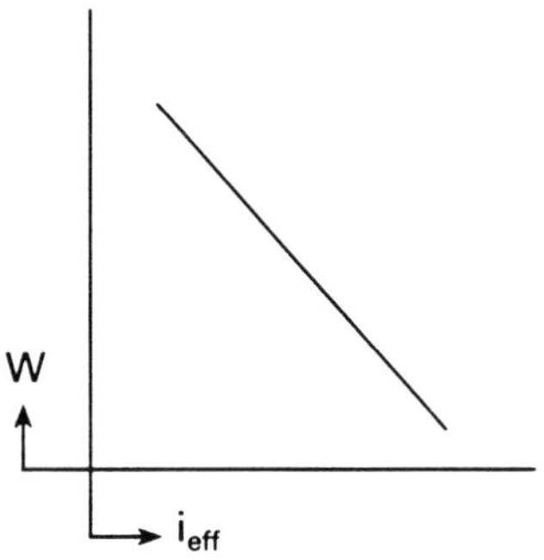

10.5 Heating, W, at material interface decreases with increasing i_{eff}.

This is not surprising considering that a high i_{eff} eliminates R_C quicker than a low current. Although a low i_{eff} results in more heating at the material interface, this does not contribute to a higher welding temperature. There is an optimum somewhere.

4 The optimum upslope time

To understand the above, a certain optimum amount of welding energy per unit resistance is assumed, i.e. $W_1 = i_{eff}{}^2 t = 2940 \times 10^3$ J/Ω. This energy is supplied to the weld with t_2 = 18 ms and t_3 = 2 ms see Fig. 10.7.

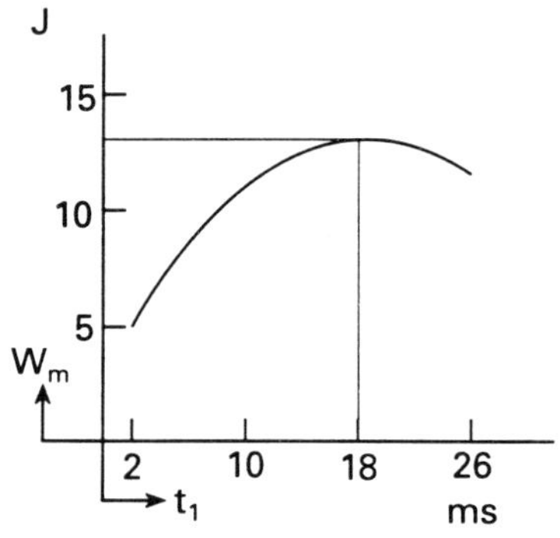

10.6 Melting heat, W_m optimum, for t_1 = 18 ms.

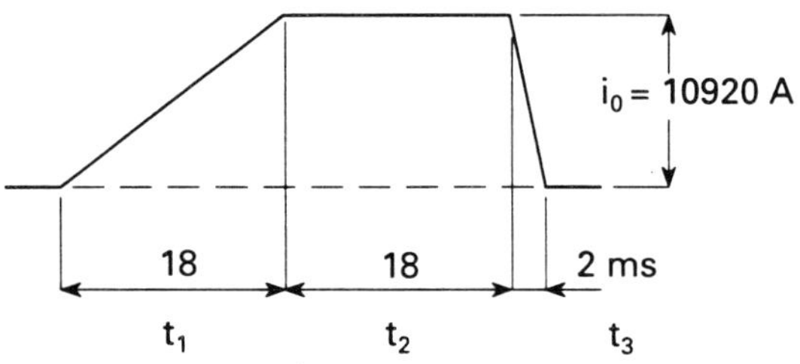

10.7 $W_1 = 2940 \times 10^3$J/Ω.

This is to ascertain the upslope time, t_1, which corresponds to the amount of welding energy, W_1, giving maximum fusion, W_m, see Fig. 10.6.

According to this graph, fusion is a maximum, when W_m = 13 J, t_1 = 18ms, thus $t_1 = t_2$. Via PHILPAC it is possible to find out what upslope time a certain amount of welding energy needs for maximum fusion.

5 Welding time

Finally, when welding energy is supplied in three periods totalling 60 ms it is less effective than when supplied in half this time i.e. 30 ms. The same temperature is reached with a 23% lower welding energy than in the case of 60ms.

These four examples show that by using the temperature calculation many problems can be analysed and discussed. Thus this method contributes to a better knowledge of the resistance welding process.

11 WELD SPLASHING

11.1 Introduction

In publications about resistance welding, splashing is dealt with mostly at the end of the weld or the peak of the welding pulse. Current density, j, becomes too large and the drop in force, ΔF_1, decreases the welding force, F_1.

The change in force, ΔF_1, is a result of too large a movable mass, excessive friction, bad welding head dynamics and electromagnetic forces.

To prevent these difficulties a 'spring compensator' in the welding head is recommended. This is specially relevant to hydraulically controlled welding heads, Fig. 11.1. The spring compensator only becomes active at the moment of fusion, activating the electrode so that it follows fast enough.

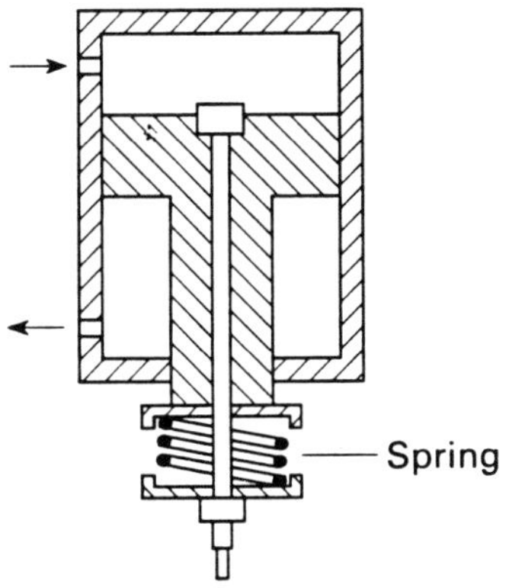

11.1 Spring compensator.

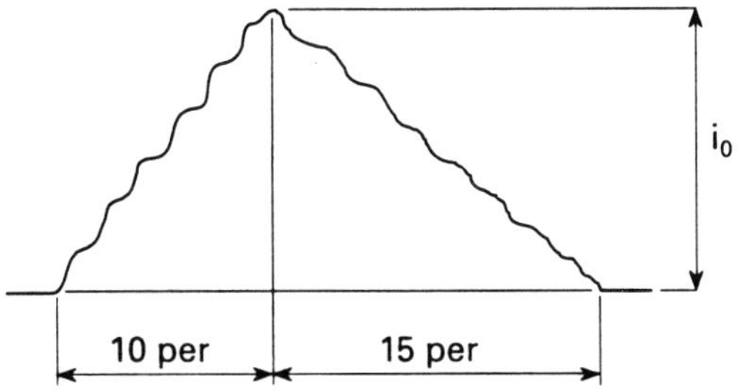

11.2 Modified pulse shape.

When splashing is a result of bad welding head dynamics, this facility is particularly appropriate.

Also it is recommended that energy is supplied gradually as a welding pulse with upslope. This applies especially when there is a layer of oxide on the part to be welded, and for anodised aluminium,[4] with which splashing may also occur at the beginning of a welding pulse.

In that case one hopes to achieve a great deal with a 3 phase rectified current pulse, see Fig. 11.2. However, this relates mainly to heavy welding. This ties in with what has already been said in the AWS handbook:[5]

> 'One of the advantages of upslope welding is its tendency to prevent splashing or spitting of metal'.

This is especially applicable for projection welding.

Normally the appearance of splashing does not seem a problem,[2] and there are not many publications dealing with it, but this is not the case where micro-welding is concerned.

In the meantime we have found that the splash symptom is a complex matter, which cannot be solved by good welding dynamics alone. Therefore we try to find out what splashing means and how it develops. Finally there are some guide-lines for splash prevention.

11.2 What are weld splashes?

First of all you have to distinguish the appearance of:

- Sparks, as a result of circuit interruption and the following gas discharge;
- Metal splashes or spitting metal parts, whereby the circuit is closed.

Sparks and arcing, recognizable by their blue colour, appear during interruption of circuit current as surrounding air is ionised. Metal splashing, however, may occur in a closed circuit when current is still flowing. Analysis of this behaviour is more complicated. In this section only the occurence of metal splashing is dealt with.

11.3 The appearance

1 Too much welding energy

The most obvious cause of weld splashing is recognizable when so much energy has been supplied that the fusion zone is greater than the area that the electrode tip can cover.

At the end of a welding pulse superfluous material is forced from between the sheets, Fig. 11.3.

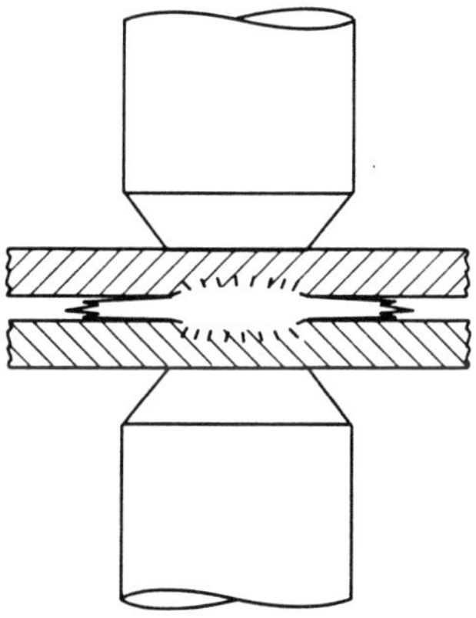

11.3 Superfluous material forced away at the sides.

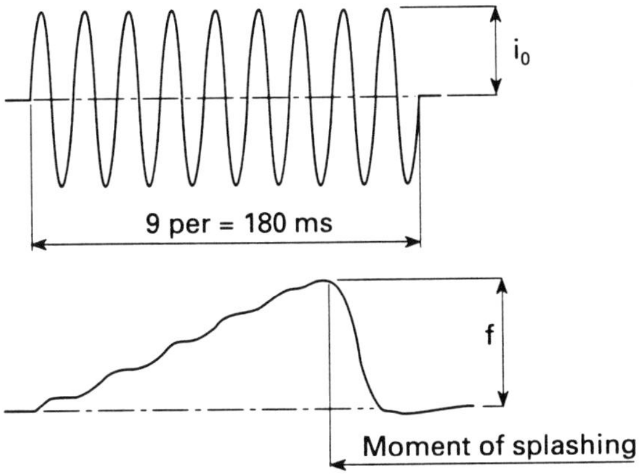

11.4 Moment of splashing identifiable by moment f.

Here splashing may spread between the two sheets. When electrodes are corroded there is also too great a heat concentration between electrode and sheet, with added risk of splashing occurring at the resulting point contacts.

2 Presence of low melting temperature

For instance, zinc, tin, lead and aluminium may produce metal vapour pressure capable of displacing the material to be welded. This type of splashing is formed from the material itself and often produces unusual shapes, Fig. 11.5.

This has nothing to do with too high an input of welding energy but with the formation of high metal vapour pressure which forces material out of the tangent plane of both sheets, and also from between electrodes and sheet.

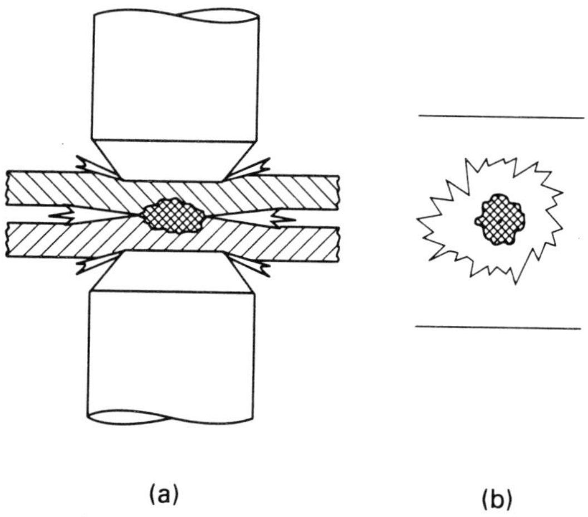

11.5 Weld splashing caused by high metal vapour pressure.

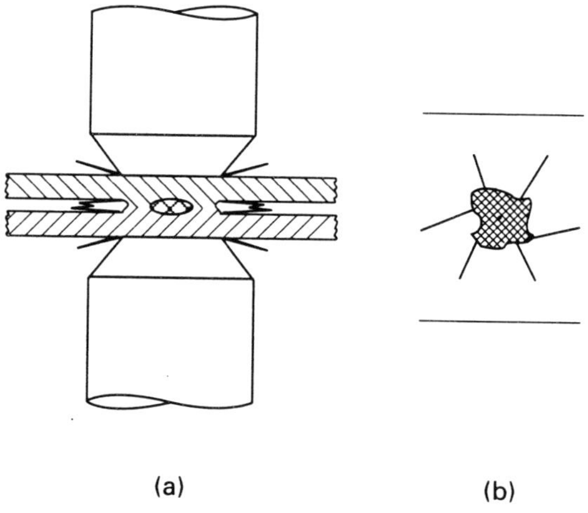

11.6 Splashing as thin formations.

In Figure 11.5b) a crown of more or less porous metal particles originates from the material between the sheets as well between electrodes and sheet.

<u>3 Oxide film</u>

The most troublesome metal splashing results from a rolling skin, or oxide films such as Al_2O_3 and Cr_2O_3. During current flow, high heat concentration occurs which is relieved by splashing of fine metal slivers.

There is hardly any electrode displacement, and splashing often starts at the beginning of a welding pulse. Measurement of welding force reveals many useful details.

<u>4 Electromagnectic forces</u>

When the welding current direction near a weld spot takes a sharp bend, a large electromagnetic field occurs at angle H, and influences the current carrying metal so that when drops appear they are forced in direction R, Fig. 11.7.

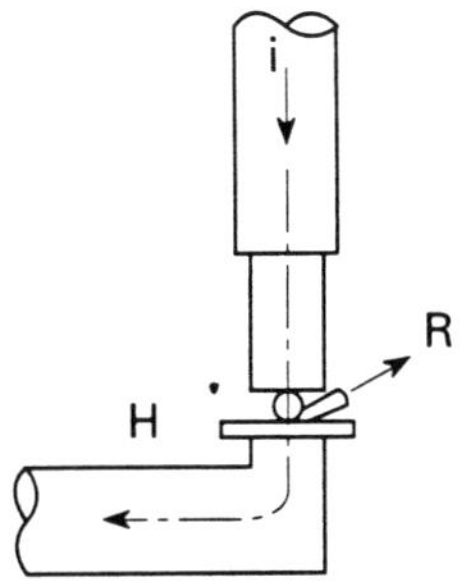

11.7 Curved current path.

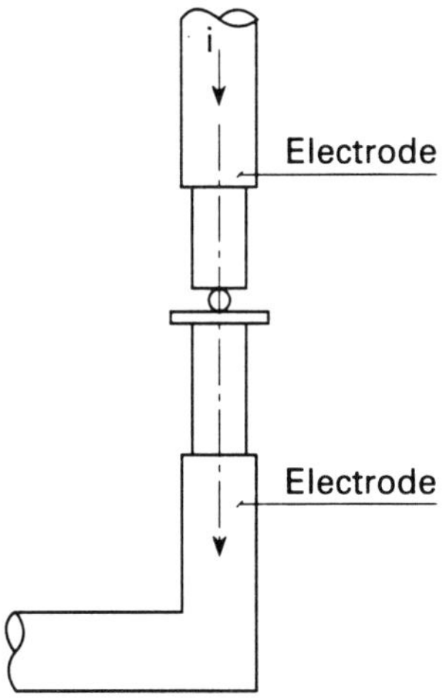

11.8 Straight current path via weld spot.

11.4 How to prevent splashing

It is clear that splash formation often possesses more features than mentioned in previous sections. To prevent splash formation it is necessary first to obtain a perfect welding head with a low movable mass, low friction and good dynamics. With this, the splash formation shown in Fig. 11.3, is easily prevented.

When the movable electrode is running well, welding force can be decreased and, consequently, also welding energy. Splash formation of the second kind, see Fig. 11.5, is controllable using an optimal working welding head.

The solution to splash problems of the third kind, splash formation as fine slivers, is use of a DC pulse with well adjusted upslope. In this way too fast an energy input is prevented. Moreover, the maximum welding pulse amplitude is practically equal to the effective value. When working with a constant current source, voltage is a guideline to splash formation.

By using voltage feedback to regulate current and power, further splash formation is prevented. With regard to the Peltier effect, see section 12, current direction also influences splashing.

Weld splashing and often metal droplets also caused by electromagnetic forces are prevented by arranging current to flow in a straight line through the weld spot, see Fig. 11.8.

Occurrence of weld metal droplets is prevented by use of a gas, so that fused material flows more easily. The right choice of material to be welded and good pre-processing both control splash formation. Use of an ultrasonic vibrating electrode is also beneficial, because it allows the current to flow more easily.

12 SEEBECK, PELTIER, AND THOMSON EFFECT

12.1 Introduction

As well as Joule heating, $W = i^2Rt$, during current flow in a resistive conductor, there is other heating from the Peltier and Thomson effects. Often these terms are used without knowing their meaning.

Their contribution is in most cases rather small, but when using direct current these effects may become important. Therefore the author thought it useful to devote a section to them, in particular the Peltier effect.

12.2 Seebeck effect

When two different metal conductors are in electrical contact, Fig. 12.1, one at temperature T_w and the other at temperature T_k, so that $T_w > T_k$, then a thermoelectrical voltage E_s occurs. E_s is proportional to the temperature difference, $\Delta T = T_w - T_k$, and this is called the Seebeck effect. When the thermoelectric potential with respect to platinum of material A is lower than in material B, $E_A < E_B$, current i flows as shown.

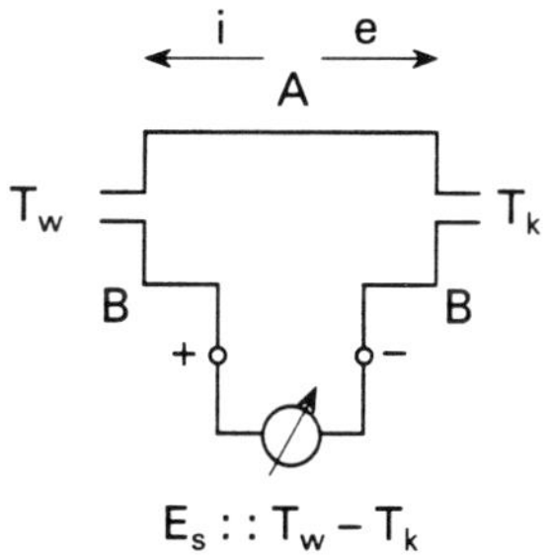

12.1 Seebeck effect, $E_B > E_A$.

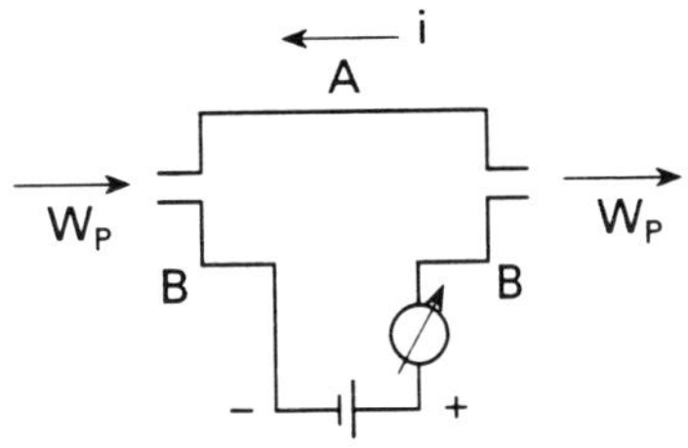

12.2 Peltier effect, $E_B > E_A$.

If both contact spots are at the same temperature and current i is caused to flow through the circuit in the direction of Fig. 12.1, an effect occurs that opposes current flow.

The right hand contact produces heat and its temperature rises, while the other spot demands heat and cools. This is the Peltier effect, Fig. 12.2.

In practical thermoelectric applications the Seebeck effect is normally used for thermoelectric generation and the Peltier effect in cooling units.

12.3 Peltier effect

1 Theory

In an atom electrons orbit its nucleus in defined paths. If energy is emitted, the electrons occupy a lower energy level than before. If temperature is reduced to a point where, except for some orbits near the nucleus, there are no longer electrons, the corresponding energy state is called the Fermi level. At this level, there is a 0.5 probability of finding an electron. The temperature at which the Fermi level occurs depends on material.

Fig. 12.3 shows:

- Metal A with a low Fermi level, N_F;
- Metal B with a high Fermi level, measured at the level where free electrons occur.

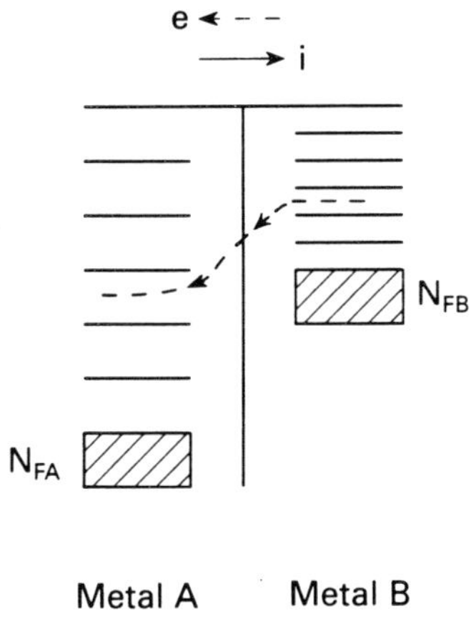

12.3 $N_{FA} < N_{FB}$.

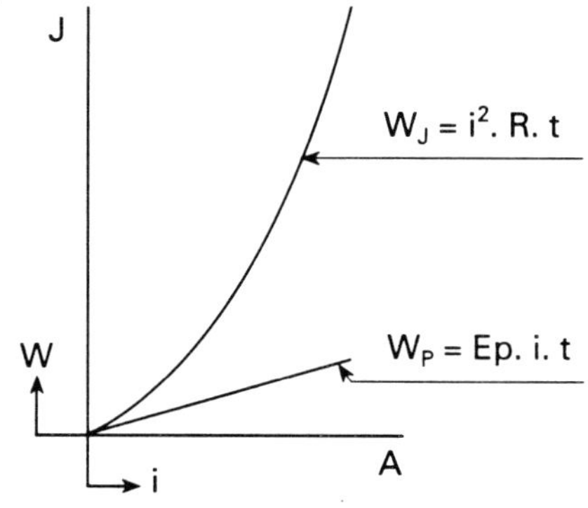

12.4 For R = E_p.

When both materials are in electrical contact with each other and a current i circulated in direction A——>B, electron flow occurs in direction B——>A. The interface temperature rises through Peltier heating.

$$W_p = E_p it$$

where

$$E_p = \text{Peltier coefficient.}$$

When the current direction is B——>A then the opposite occurs as electrons go from a low to high energy level, and heat has to be supplied. This is abstracted from the surroundings and the interface cools. Electrons move from a high energy level to a lower level which means that energy is freely available.

2 Thermal management

Peltier heating is expressed by

$$dW_p = E_p idt$$

where

E_p = Peltier coefficient

$$= T \frac{dV}{dT}$$

T = temperature of the contact interface, K,

V = thermal voltage.

In comparison with Joule heating, Peltier heating is proportional to i and dependent on the current direction. When both sources of heating are plotted against current i, the result is the graph of Fig. 12.4. This shows that, especially at the beginning of the current pulse, the Peltier effect is very important. As i increases, W_J dominates. In microwelding the first part of the pulse is often responsible for the final stage of the welding process.

Moreover, heat production depends on the coefficient, E_p, which has the dimension of voltage.

When current flows through two different metals in electrical contact, a voltage difference ΔV occurs, called the thermoelectric voltage. In Fig. 12.5 this ΔV for various metals with respect to platinum is plotted as a function of temperature T, see graph and Table 12.1.

Table 12.1 Thermal electromotive force (millivolts) of elements relative to platinum*

	Temperature of hot junction °C			
	- 200	- 100	+ 100	+ 200
Aluminium	+ 0.45	- 0.06	+ 0.42	+ 1.06
Antimony	---	---	+ 4.89	+ 10.14
Bismuth	+ 12.39	+ 7.54	- 7.34	- 13.57
Cadmium	- 0.04	- 0.31	+ 0.91	+ 2.32
Caesium	+ 0.22	- 0.13	---	---
Calcium	---	---	- 0.51	- 1.13
Carbon	---	---	+ 0.70	+ 1.54
Cerium	---	---	+ 1.14	+ 2.46
Cobalt	---	---	- 1.33	- 3.08
Copper	- 0.19	- 0.37	+ 0.76	+ 1.83
Germanium	- 46.00	- 26.62	+ 33.9	+ 72.4
Gold	- 0.21	- 0.39	+ 0.74	+ 1.77
Indium	---	---	+ 0.69	---
Iridium	- 0.25	- 0.35	+ 0.65	+ 1.49
Iron	- 3.10	- 1.94	+ 1.98	+ 3.69
Lead	+ 0.24	- 0.13	+ 0.44	+ 1.09
Lithium	- 1.12	- 1.00	+ 1.82	---
Magnesium	+ 0.31	- 0.09	+ 0.44	+ 1.10
Mercury	---	---	+ 0.06	+ 0.13
Molybdenum	---	---	+ 1.45	+ 3.19
Nickel	+ 2.28	+ 1.22	- 1.48	- 3.10
Palladium	+ 0.81	+ 0.48	- 0.57	- 1.23
Potassium	+ 1.61	+ 0.78	---	---
Rhodium	- 0.20	- 0.34	+ 0.70	+ 1.61
Rubidium	+ 1.09	+ 0.46	---	---
Silicon	+ 63.13	+ 37.17	- 41.56	- 80.5
Silver	- 0.21	- 0.39	+ 0.74	+ 1.77
Sodium	+ 1.00	+ 0.29	---	---
Tantalum	+ 0.21	- 0.10	+ 0.33	+ 0.93
Thalium	---	---	+ 0.58	+ 1.30
Thorium	---	---	- 0.13	- 0.26
Tin	+ 0.26	- 0.12	+ 0.42	+ 1.07
Tungsten	+ 0.43	- 0.15	+ 1.12	+ 2.62
Zinc	- 0.07	- 0.33	+ 0.76	+ 1.89

Comment: A positive sign means that in a simple thermoelectric circuit the element is positive to the platinum at the reference junction (0 °C).

* Based on tables in 'Temperature', New York: 1941. The values below 0 °C were not usually determined on the same samples as the values above 0 °C.

The emf generated by any two elements, A and B, can also be found from this table. It is the algebraic difference, $A_e - B_e$, between the values (A_e and B_e) for the emf generated by each relative to platinum; a positive sign indicates that A is positive to B at the reference junction, and a negative sign that A is negative to B.

This graph and table give an impression of what can be expected from thermal management for current flow through metallic couples. Here are some practical examples.

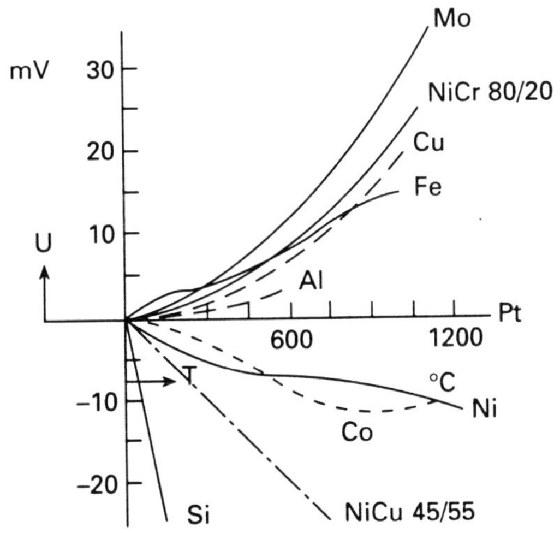

12.5 V = f(T) T in C
V with respect to platinum.

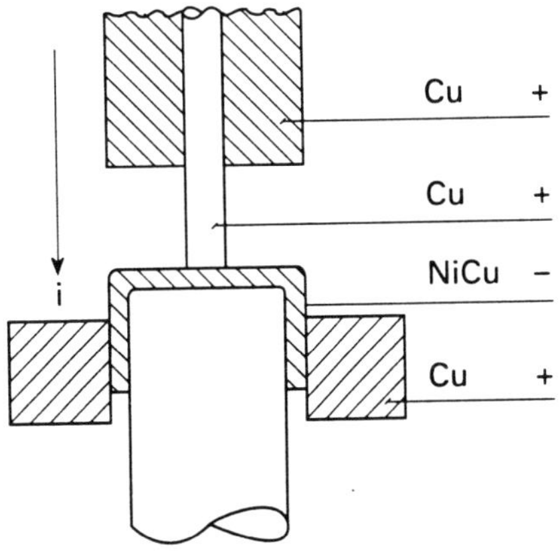

12.6 Practical example 1.

3 Practical examples

i When tinned Cu wire of 0.7 mm diameter is welded to a sheath of NiCu, see Fig. 12.6, with direct current then current flow from wire to sheath is required. Figure 12.5 shows that for current flow from wire to sheath an E_p acts, that adds to Joule heating, whereas from cover to electrode cooling occurs. The wrong current direction caused problems when welding with a capacitor discharge process.

ii An earlier example is mentioned in reference 1.
Two Mo-wires of 0.1 mm diameter were welded to Ni wire of 0.75 mm diameter, Fig. 12.7. When the correct current direction was used

for the upper connection, i.e. current flowing from above to below, the lower weld was supplied with insufficient energy. When the current direction was reversed the upper weld failed.

iii A recent example is in Fig. 12.8.
Two sheets were welded with chromium copper electrodes at $i = 950$ A and $t = 25/10/5$ ms.
Here,

$$E_{p\,Fe} > E_{p\,Cu}.$$

For a given current direction there is, at a Cu——>Fe interface, Peltier cooling and at an Fe——>Cu (of the electrode) interface Peltier heating. As a consequence the weld nugget moves in the direction of the current. At current reversal the weld nugget moves to the top. Note that a rather long upslope time is used, i.e. t_2 = 25 ms.

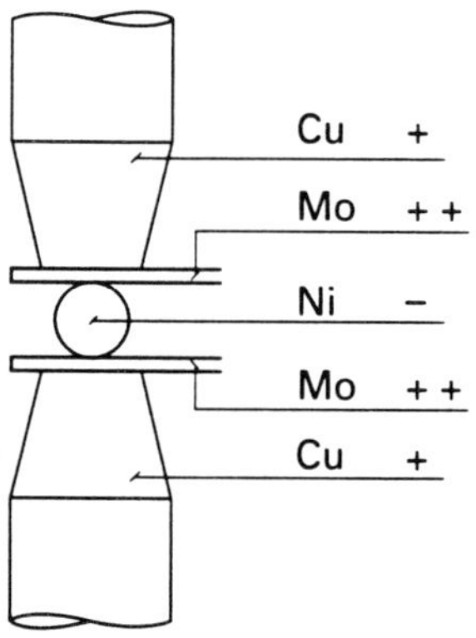

12.7 Practical example 2.

iv Worked example
Take example C2 as a starting point
Peltier heat

$$dW_p = iE_p dt$$

$$dW_p = iT \frac{dV}{dT} dt$$

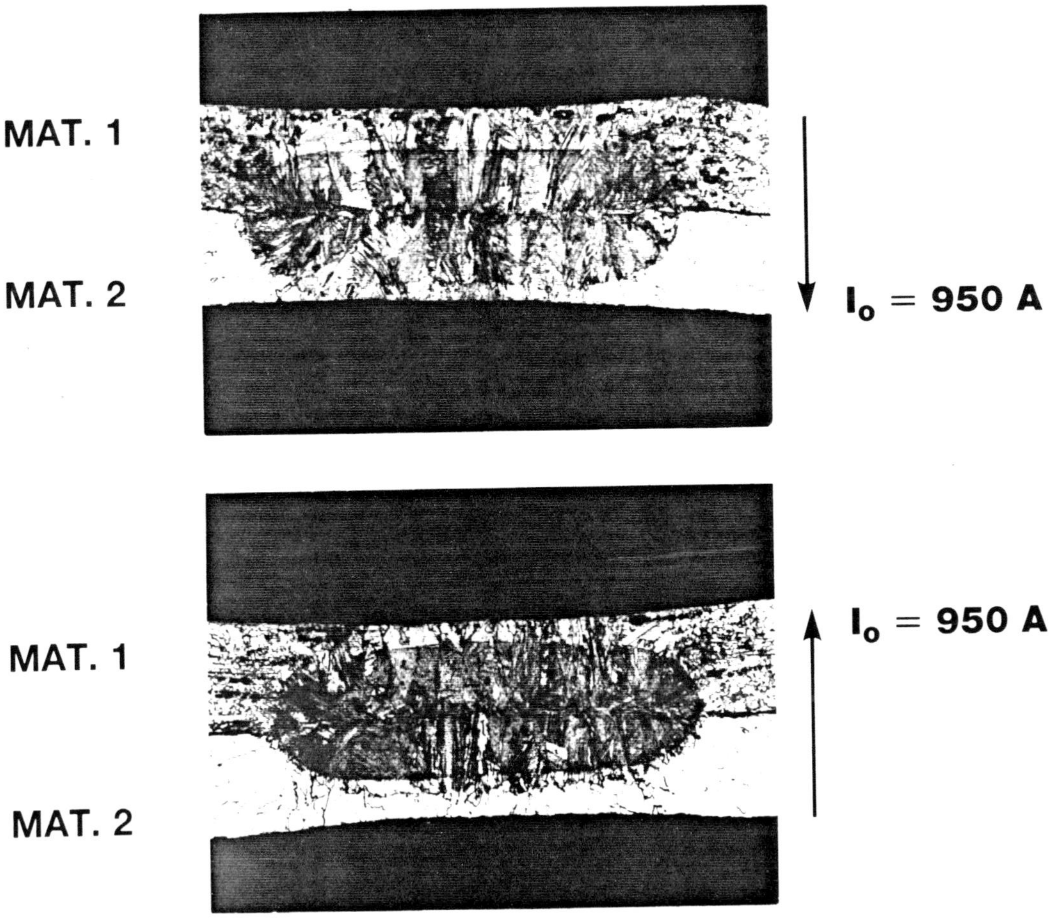

12.8 Practical example 3.

Imagine that welding was done with a half cycle idt = $i_{average}$ = 300 A and t = 10ms.

From Table 12.1 we find

for Mo $\frac{dV_{Mo}}{dT} = \frac{3.19}{200}$ mV/C

for Ni $\frac{dV_{Ni}}{dT} = -\frac{3.1}{200}$ mV/C

From this $\frac{dV}{dT} = \frac{3.19 + 3.1}{200} = \frac{6.29}{200}$ mV/C

Assume that interface temperature between Mo wire and Ni wire is 1400 C.

then $T \frac{dV}{dT} = \frac{6.29}{200}$ 1400 = 44 mV

For $i_{average}$ = 300 A there is Peltier heating

$$W_p = 300 \frac{44}{10^3} \frac{10}{10^3} J$$

$$W_p = 0.13 J$$

Peltier heat is supplied to one part of the weld and withdrawn from the other for a given current direction. It is recommended to weld such a joint with a current pulse of 1000 Hz to counteract uneven heating.

12.4 Thomson effect

1 Theory

The Thomson effect is the electromotive force (voltage) produced by temperature difference in a single conductor and the heat change associated with current flow between temperature differences.

As current carriers, for instance electrons, in metal have an electric charge and transport heat, the total action is thermoelectrical.[5]

When, in addition to this, the temperature gradient $\frac{dT}{dx}$ exists, the energy carriers have a tendency to move from hot—> cold, Fig. 12.9.

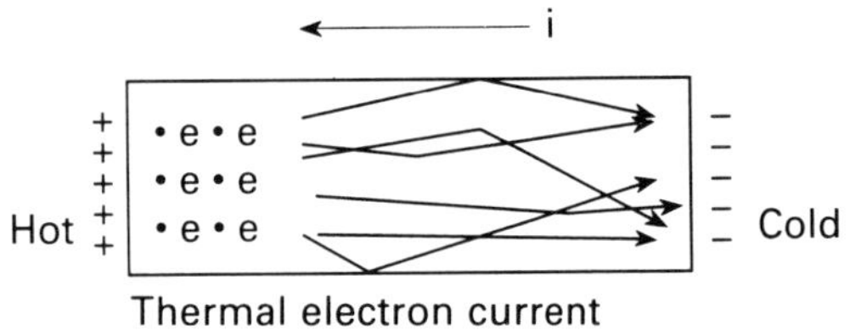

12.9 Thermoelectrical capacity in a metal bar with temperature gradient.

As a result of this effect the ends of the conductor become electrically charged. There is a current, i, from cold to hot, and therefore opposite to the temperature gradient. The voltage drop with change in temperature is the thermoelectric capacity, S,

Shown as $S = \frac{dV}{dt}$, where

V = voltage

T = temperature

For materials where charge carriers are electrons $S < 0$.

For materials where charge carriers are holes $S > 0$.

2 Thomson effect

When electrons move from a hot to a cold region they supply heat to maintain thermal equilibrium with the surroundings, which results in an effect called Thomson heating.

When, on the contrary, electrons flow from cold to hot they demand extra energy from the surroundings, resulting in cooling.

When the Thomson energy $= W_{Th}$

then for current i during time t:

$$dW_{Th} = \delta \, it \, \frac{dT}{dx} \, dx$$

Thomson coefficient $\delta = T \frac{dS}{dT} = T \frac{d^2V}{dT^2}$

Coefficient has dimensions $\frac{V}{T}$.

Note that the Thomson effect is proportional to current i, temperature gradient $\frac{dT}{dx}$ and voltage gradient $\frac{dV}{dx}$.

Here just one material is concerned. Lander[2] measured and calculated the Thomson coefficient of various metals, see Fig. 12.10. This figure shows that only $\frac{1}{30}$ μV/K is involved, so that the Thomson effect in resistance welding is not an important factor.

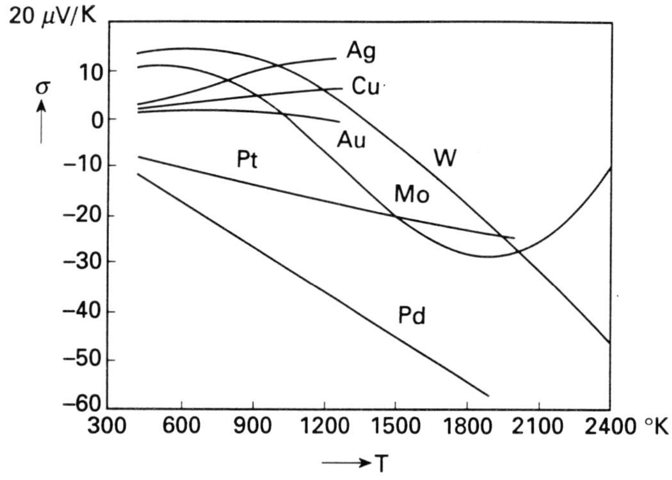

12.10 Thomson coefficient. When $\delta > 0$ heat is evolved when dt/dx and i have opposite directions.[2,3]

12.5 Conclusion

Although many materials do not differ much regarding thermo-electrical voltage it is advisable, especially when direct current is used, to pay attention to the possible influence of the Peltier effect.

This consideration is of great importance as, right at the beginning of a welding pulse, the final result of the process has already been determined.

The influence of the Thomson effect on resistance welding can be considered negligible.

13 WELDING EQUIPMENT

13.1 Introduction

Joule heating in resistance welding for current flow through a resistive workpiece is $W_1 = i^2Rt$.

R is a dependent variable and cannot be adjusted. The i and t terms on the contrary can be adjusted. In particular, i is an important parameter as it occurs in W_1 quadratically. To operate a well controlled welding process it is necessary to adjust parameters i and t as precisely as possible.

It is surprising how little attention has been paid to these obvious factors until now. Adjustment of welding machines is rather limited, for instance, those which operate on alternating current at 50 Hz, see Fig. 13. 1a) and Fig. 4.1. The amplitude i_o is adjustable in steps and for values in between phase shift is used, Fig. 13.1b).

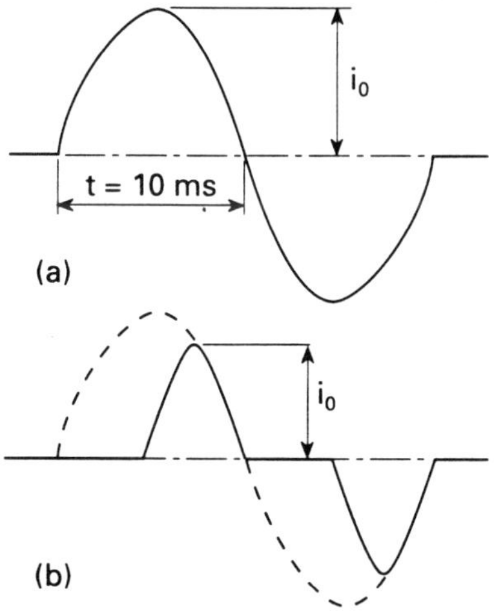

13.1 Amplitude adjustment via phase shift.

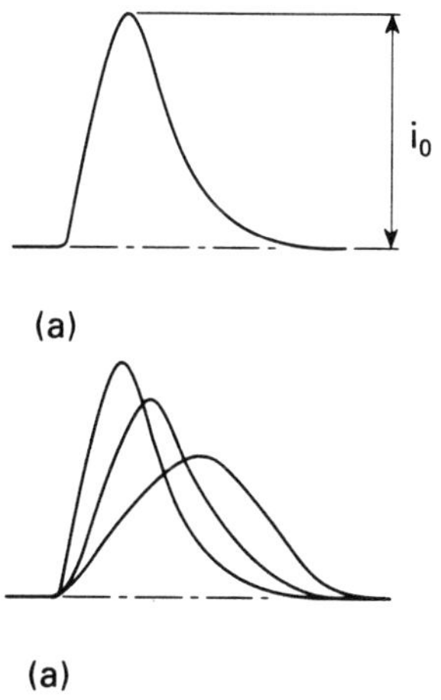

13.2 RC adjustment.

When heavy welding equipment is used one has to rely completely on phase shifting. Welding energy supplied in this way cannot be defined and measured easily.

Moreover the time between the welding command and start of the welding pulse is variable, and there are difficulties at higher frequencies.

For light welding with a higher frequency than 3 Hz, a transformer with capacitor discharge is used, see Fig. 13.3.

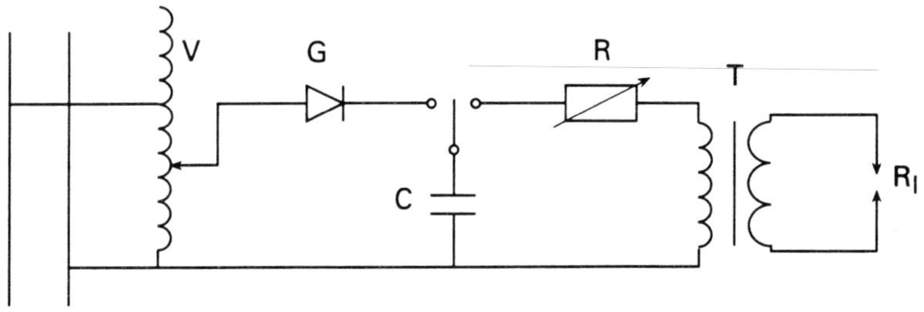

13.3 Transformer-capacitor (traco) method.

But even for this welding process the range of adjustment is limited. Current pulse amplitude is controlled by the DC supply voltage and pulse shape is affected by the RC network and circuit inductance, Fig. 13.2.

13.2 New developments

Taking into account the previous comments one looked for an electrical power supply with adjustment facilities that meet more efficiently requirements imposed by precision, volume production and stability. Need for these is emphasised by the introduction of automatic welding machines with high welding frequencies of 10-25 welding pulses per second.

At first there were many experiments with welding pulses, using 2000 Hz and later 1000 Hz.

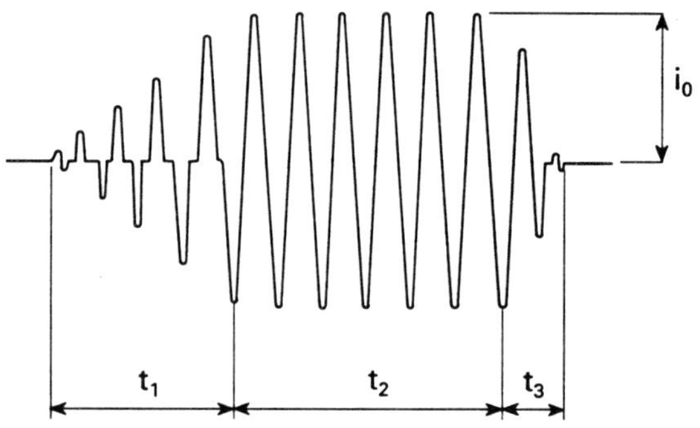

13.4 1000 Hz welding pulse.

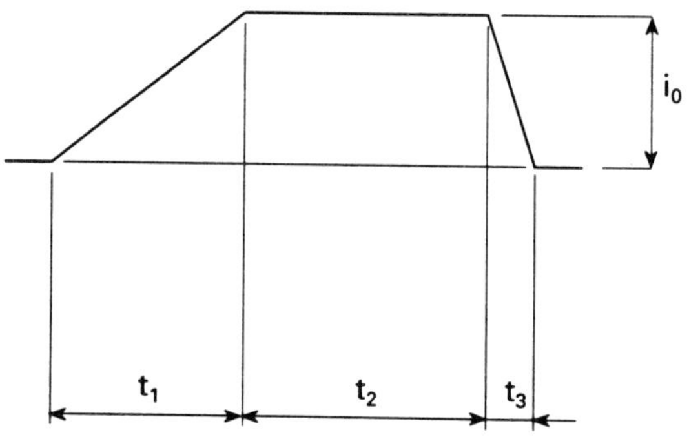

13.5 DC pulse.

The amplitude i_o and the welding times t_1, t_2 and t_3 of these welding pulses are adjustable separately. After co-operation with a welding machine manufacturer, welding equipment has become available for welding pulses of i_o = 1000 A and higher. This has led to considerable improvement of the adjustment possibilities.

However objections have remained regarding this method. High impedancy in the secondary circuit occurs as a result of high frequency of the alternating current. The current cables must stay short and be fixed very carefully, for instance coaxial. Moreover when operating with a welding pulse of some milliseconds, the energy supply will again become undefinable.

Finally, equipment has been developed that supplies pure direct current, Fig. 13.5. The term 'direct current' will be used in the text as it is more commonly used and recognized than the correct term for short welding pulses, 'monopolar pulse'.

There are new possibilities regarding the control of welding pulses. The whole process can be monitored via a computer program and controlled during the welding cycle.

13.3 DC welding pulse advantages

Apart from some disadvantages already mentioned, the advantages outlined in this paragraph also apply to 1000 Hz operation. Now it is possible to adjust the welding pulse precisely,but what tells us with certainty that the adjustment is correct? In the first place this knowledge is provided by practical experience in the laboratory and factory. Because of the established sale of this equipment within the Philips concern, one concludes that the client is satisfied. By thermal management calculation, see section 10, various factors affecting welding are analysed.

Experiments show the following:

- Welding energy, even when using short welding times and with a continuous heat sink, is measurable and calculable.

- At the instant the welding command is given the pulse appears, which is of great importance at high welding frequencies.

- Welding is controlled using closed loop feedback. For instance, when movement has reached a certain value, current stops, or when voltage increases too quickly the weld is rejected to prevent an eventual splashing.

- Beside the DC unit adjustment as a current source, i_o = constant, it is possible to keep the voltage e = constant, often required for parallel gap welding. Finally the unit can be set to operate with ei = constant.

- By correct choice of upslope gradient, the welding head is lightly loaded compared with a steep upslope.

- For materials sensitive to polarity because of the Peltier effect (see section 12,) such as NiCu, Ni, Mo, Al, etc, current direction has a decisive effect.

- Experiments with crosswire welding of different materials with DC, 1000 Hz and 50 Hz show that the adjustment range for good welding with DC was considerably larger than with 1000 Hz and 50 Hz.

- Also, feedback of measurable process factors via a computer program is effected satisfactorily.

- By applying a second pulse after the welding pulse, material sensitive to heat, such as spring steel, is tempered so that the joint remains flexible.

- Operational life of electrodes generally increases, as high current peaks do not occur.

- Weld splashing is better controlled with DC than AC.

When comparing results obtained using DC with those of AC, especially at 1000 Hz, remember that DC magnetises ferromagnetic materials and the weld area should therefore be non-magnetic. The splashing problem is reduced but use of DC does not yet signify splashless conditions, see section 11.

When providing the DC unit with a power control adjustment ei = constant, it seems that splash control also offers new possibilities.

13.4 Equipment

1 50 Hz AC equipment

Before the DC era, the Resistance Welding Group designed AC microwelding machines. The advantage in comparison with existing welding equipment was that the welding current amplitude could be controlled continuously by means of a variac (variable transformer). As the variac determines load limits this method is not suitable for heavy welding work and welding current supplied by a variac should not exceed 3000 A. A transformer with sufficient output voltage steps and accurate phase shift control provides good adjustment also.

The maximum current of the microprocessor controlled AC welding power supply, Fig. 13.6, is about 15 000 A. The secondary voltage can be adjusted in nine steps.

Using a measuring resistance, Fig. 13.8, the welding pulse can be displayed visually on a storage oscilloscope which ensures a consistent welding current measurement. For higher currents it is better to use a contactless current monitoring system, see Fig. 13.7.

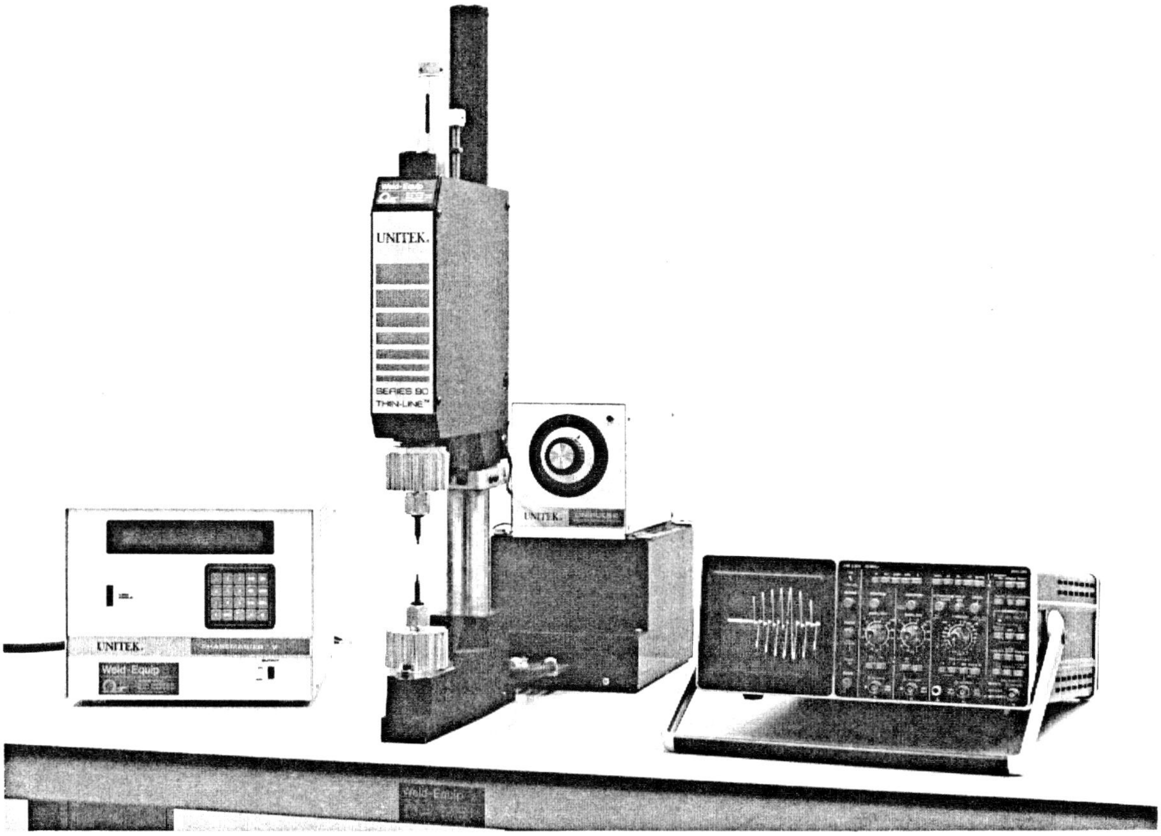

13.6 AC welding system.

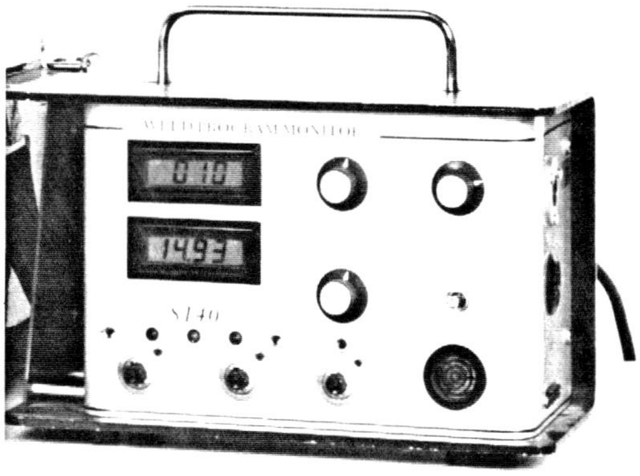

13.7 Current monitor.

13.8 Low inductance shunt for current measurement.

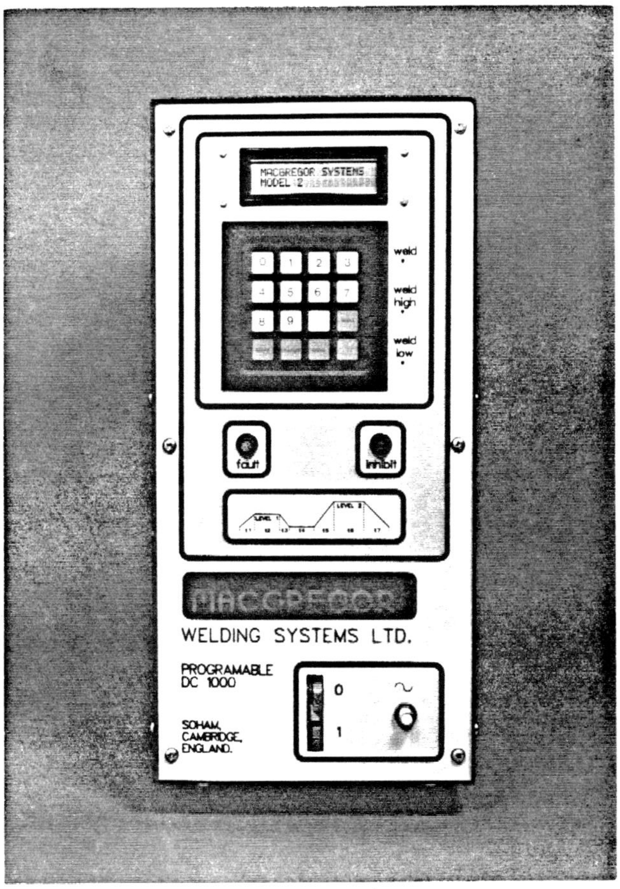

13.9 1000 Hz supply, Type 2W.

2 DC equipment i

Our Resistance Welding Group has been busy for many years with direct current research. The main problem seems to be manufacture of equipment with sales potential. Finally the group has found in MacGregor Welding Systems,* Mildenhall, England a willing supplier. These contacts have resulted in launching the DC standard power unit on the market.

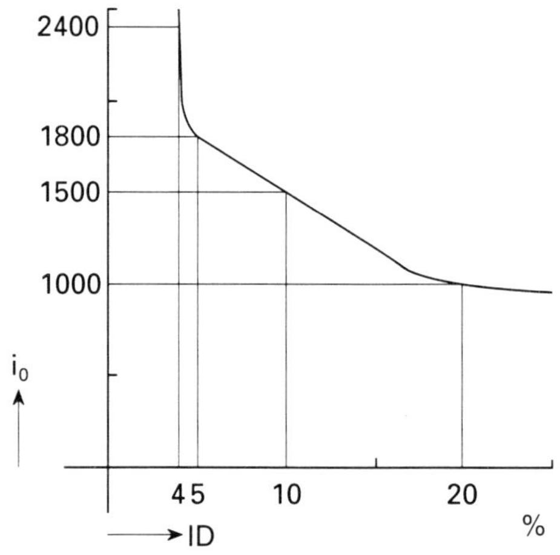

13.10 Duty cycle of welding current.

Technical data:

- Practical range of i_o = 1000 A, 1800 A, 2400 A, 3500 A, 5000 A, 10 000 A with a duty cycle according to Fig. 13.10;
- Suitable for i = constant, e = constant and ei = constant;
- Open circuit voltage asjustable to 4 and 6V;
- Air cooled;
- Double welding pulse and facility for connection to a computer;
- t_{max} = 99/99/99 ms;

A block diagram is shown in Fig. 13.11. As shown in Fig. 13.12 a single unit can be set, using multiplexing, to drive many different workstations, each with its own control.

* Representative for MacGregor, Mildenhall, UK, is Weld-Equip BV, Engelseweg 217, Helmond, Holland.

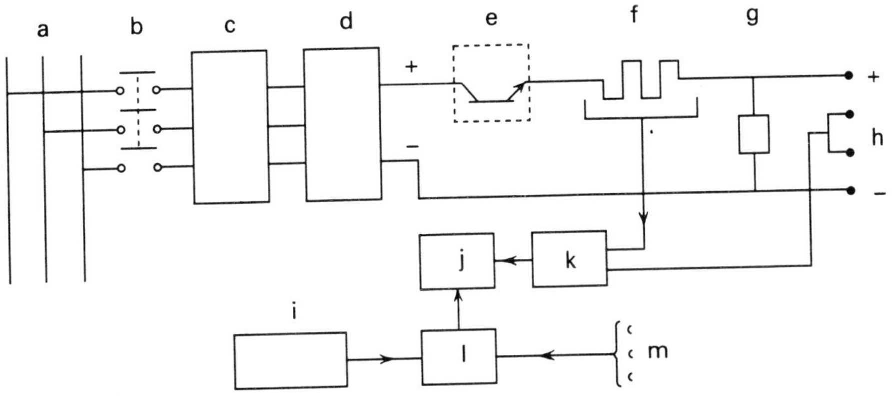

13.11 Block diagram of the transistor-controlled DC sources.

a Three phase power supply
b On/off contactor
c Welding transformer
d Rectifier
e Transistor control unit
f Feedback shunt
g Transient protection devices
h Voltage sensing
i Welding adjustment unit
j Control unit
k V or I feedback
l Pulse shape generator
m External control inputs

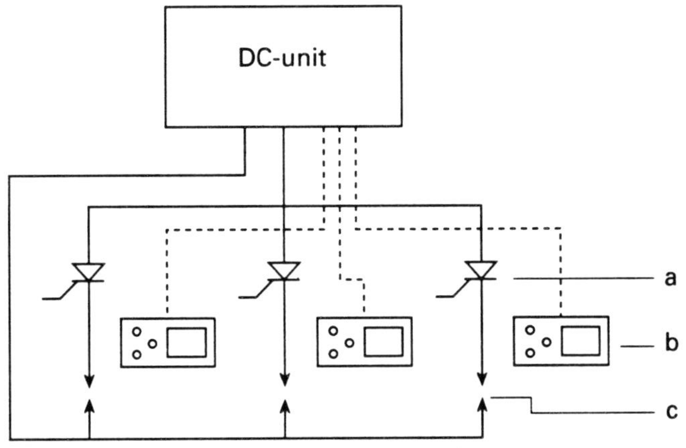

13.12 A centrally positioned power supply with individual current and time settings for each welding station. adjustment facilities.

a Distribution unit with thyristor switch
b Welding adjustment
c Welding head

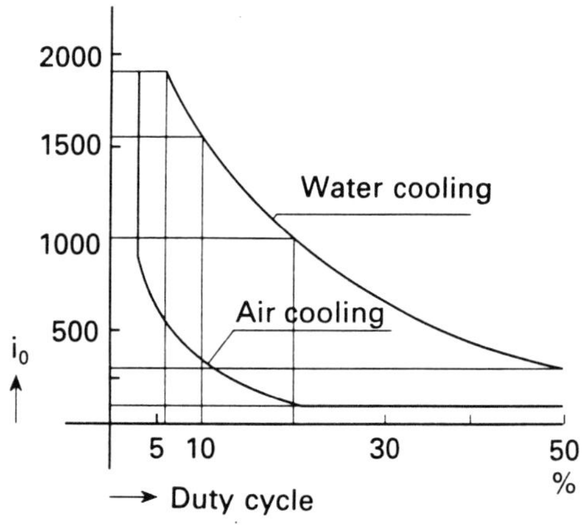

13.13 Welding current duty cycle.

14 PROCESS CONTROL

14.1 Introduction

When one mentions process control, high expectations are raised and one imagines sophisticated systems that guarantee optimum weld quality despite variations during welding.

One should realize, however, that process control means, in the first place, adequate adjustment and control of various welding parameters. It is only when one or more factors relate directly to weld quality that quality control can be achieved. Which parameter-quality relationship is concerned depends on the intended result, see, for instance, section 4.5.

Determining this relationship is a question of, among others, research, experience and metallurgical knowledge. This section deals mainly with process control.

Moreover, it is useful to point out another contrast. Detailed process control over many inherently variable factors is one way to achieve reproducible welding results. Another approach is to use welding equipment with built-in stability, such as i_0 or e or ei = constant etc, so that special external process control is hardly necessary. The examples in section 4.5 are of process control achieved without complicated modifications. The DC power supply unit in particular was taken into consideration.

Processes operating via feedback control are often vulnerable and welding adjustment more difficult than for processes without these special facilities. It is useful to examine various control methods such as use of computer programs, regarding welding inputs as energy, charge, or force, by measurement of, for instance, electrode voltage with the aim of controlling current and time during welding.

The final requirement is welding equipment with stable characteristics and producing the necessary weld quality, see also section 3.

14.2 Welding factors

During resistance welding there are many factors which influence the process and hence the weld. These factors and their occurrence during a welding cycle are classified in three stages, Fig. 14.1.

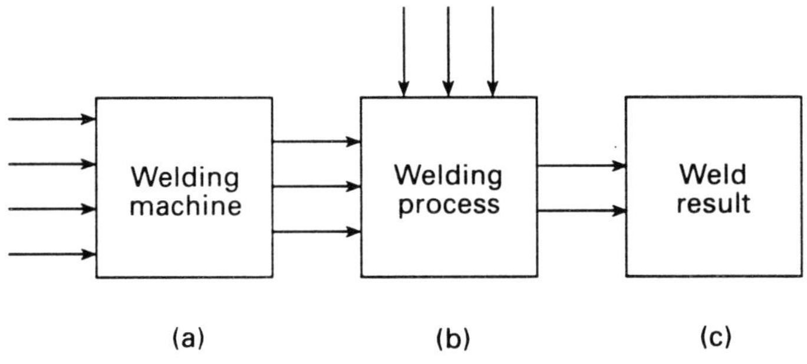

14.1 Three welding cycle stages.

<u>Stage a, the welding machine</u>

Here the parts to be joined are supplied, positioned, placed in the clamps and welded to a specified strength. Electrical power is supplied, current flow and welding time adjusted, then the welded product is removed. During this stage anything can be done to influence positively the final welding cycle; for example, correct time, current adjustment and good positioning.

Stage b, the welding process

Here factors such as thermal management, weld resistance, electrode wear, splashing, etc are important. These are not adjustable, as the process develops because of what was determined in stage a. Also external factors may be present which cannot easily be controlled, such as ambient temperature, dampness, vibration, etc. What can be monitored during welding through measurement are current, voltage, resistance, energy, movement and welding force.

By making such measurements the welding process is monitored and ultimately controlled. This is the stage at which process control action can be applied. Only how to intervene and what to control determine the following stage.

Stage c, weld result

This is the product of what preceded in stage b, and must give acceptable weld quality and consistent reproducibility, see also section 3.3. With these requirements as a guide the researcher should measure, monitor and, if necessary, control certain parameters during stage a. This demands research, physical knowledge, also good measuring equipment and techniques. Welding process control depends on the available welding equipment. In this case a DC power supply provides the necessary control facilities.

Classification in three stages should give a clear view of the various factors that influence and determine the welding result. This classification is, however, capable of further improvement.

14.3 Terminology

Already mentioned are measurement, monitoring and control.To make clear what is meant by these process control terms, see Fig. 14.2.

(a) Uncontrolled process — No adjustment possible. Example: a waterfall.

(b) 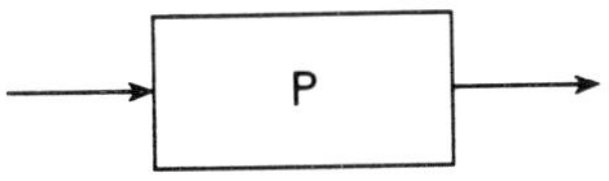

Adjustment of process p. Example: water tap.

(c) 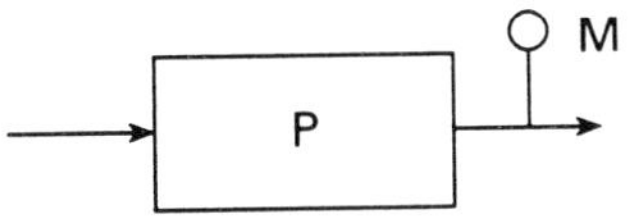

Monitoring. Example: water tap with pressure gauge.

(d) 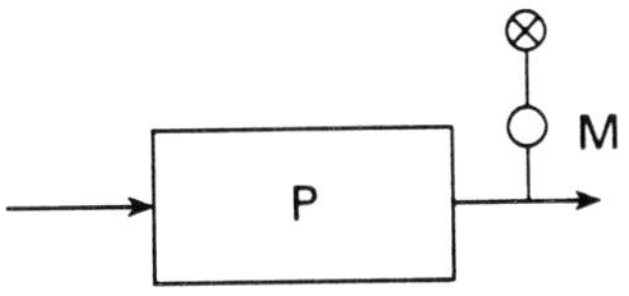

Monitoring and error detection. Example: indicator lamp but no process intervention.

(e) 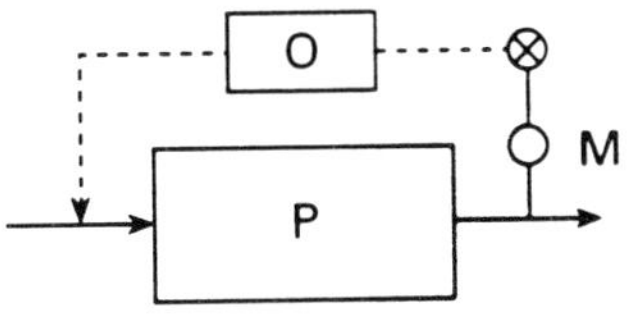

Open loop process control. Process control via operator, thus automatic.

(f) 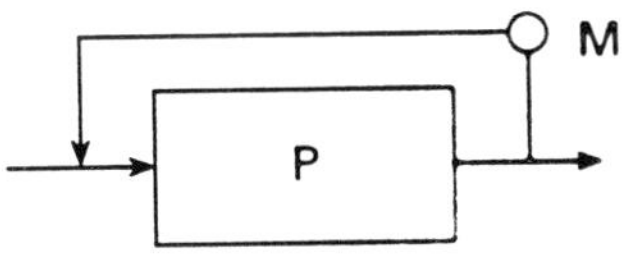

Closed loop process control. Process regulation via feedback loop, measurement and essential automatic control.

14.2 Process control.

Figure 14.2 shows that for process P, welding equipment L can be substituted, where the power unit is equipped with a microcomputer as controller.

Some examples of practical applications are given below to give an impression of the possibilities.

14.4 Process control examples

When looking for examples of process control in resistance welding one realizes that there are few, partly because there is not much demand for process control and/or it is often difficult to obtain. Use of DC equipment opens up, without doubt, new possibilities. Below are some practical examples from our company and the welding world in general.

1 Differing from most common welding installations using constant current is the bridging of gaps between printed circuit board conductors using constant voltage. As temperature rises in the strip between the electrodes, resistance R increases as well.
With e = constant, heat build-up $W = \frac{e^2}{R} t$ decreases.

This is an example of simple process control under category 14.3 b) see Fig. 14.3.

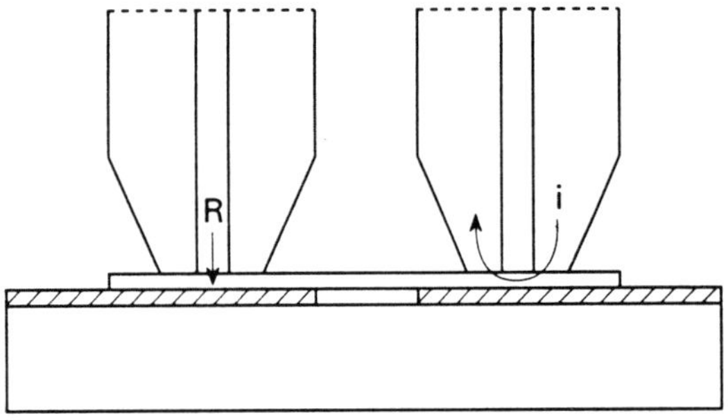

14.3 Simple process control.

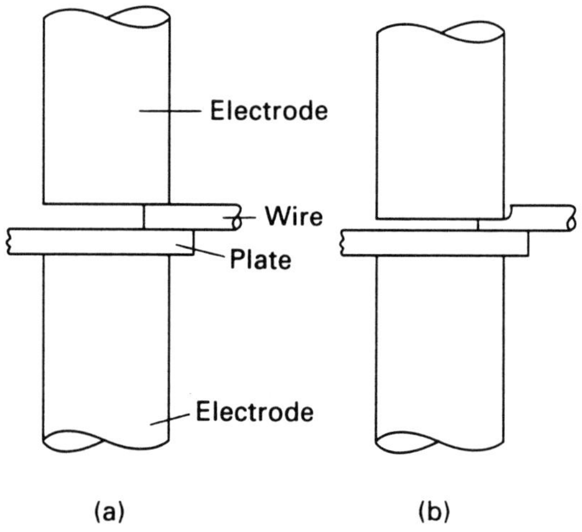

14.4 Process control.

2 A weld has a lower resistance in a wire to sheet joint if there is insufficient wire in contact with the electrode during welding. This is different in the case of a wire that does pass through sufficiently far, Fig. 14.4. With i = constant, welding voltage drops to a specified limit. This is followed by an out of tolerance signal and the faulty weld is identified. This is called monitoring and fault detection, see category 14.3d).

3 If, during welding of oxidised wire to sheet the voltage across the weld exceeds a certain value and i = constant, there is a risk that splashing may occur, Fig. 14.5. One can terminate welding via feedback after time t_1, or switch off the current pulse at a lower level. In either case, the occurrence of splashing is prevented and electrode wear reduced.

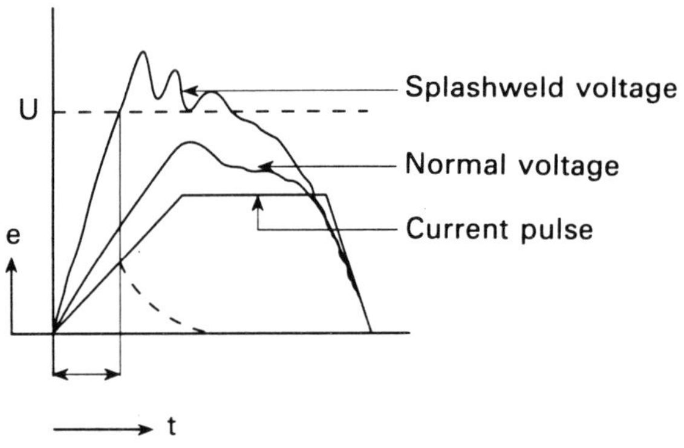

14.5 Voltage and splashing relationship.

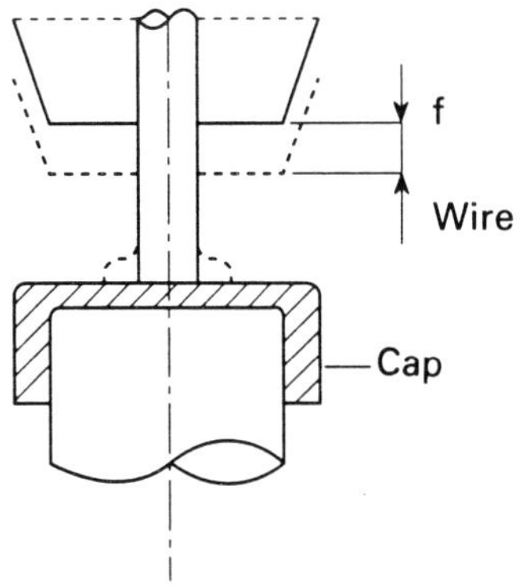

14.6 Wire to cap and movement of f.

4 When welding a wire to a sheet or a cap vertically, Fig. 14.6, movement f can be used as a quality measurement criterion. When f reaches a certain value the current supply is switched off. These two examples rely on welding pulse regulation via feedback and belong, therefore, to the process control category 14.3f).

5 When two sheets are spotwelded with a machine that works at 50 Hz, voltage across the weld drops when the weld nugget reaches a certain size. When voltage falls below a specified point, welding current is turned off, Fig. 14.7. This principle is used by the ARO Machinery Co, London, as a process control system type called Spotrite.

6 The starting point in spotwelding is expansion of sheets to be welded when current flows, Fig. 14.8. When sheet expansion reaches a certain value, it is assumed that the weld nugget is sufficiently large, at which point current is switched off. It is not known by the author whether this system is in practical use.

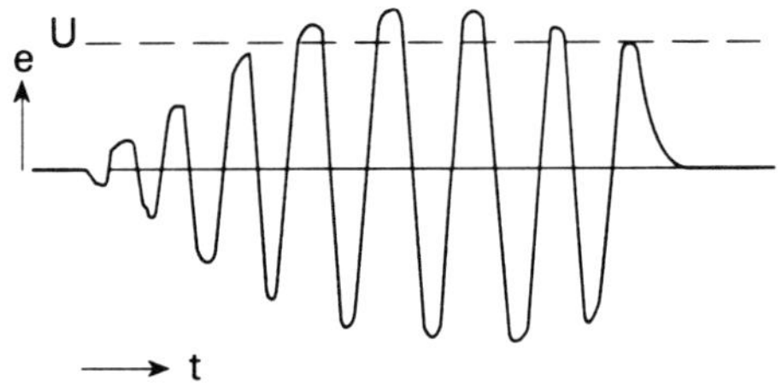

14.7 Welding pulse is terminated when e drops to U.

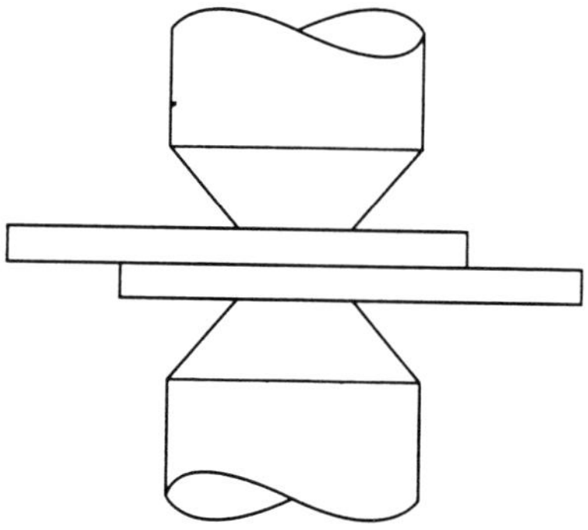

14.8 Process control via sheet expansion.

The welding process in the last four examples is controlled by terminating the energy supply. A DC power supply unit, however, permits more accurate adjustment of the welding process.

15 TROUBLESHOOTING

15.1 Introduction

To make a good judgement it is necessary to distinguish between actual failures and welding problems. Welding problems refers to all aspects of the resistance welding technique. Failures refers to cases that deviate from specified welding conditions.

There are failures related to the welding techniques used, but also failures that have no relationship with the welding process. After the product has been welded, one can check for possible failures. First of all one has to check that welding power and welding adjustment are correct, and especially whether the welding machine is operating correctly. An example is given below. Operational life of electrodes used in a welding machine was 35 000 welds. When different equipment was installed electrode operational life dropped to 5000 welds. The welding process was unsatisfactory, but detailed analysis proved that in the machine the welding force was not being transmitted sufficiently to the material to be welded. The effective welding force was much too low.

The following section gives guidelines regarding use of resistance welding and explains possible failures.

15.2 Trouble sources

1 Mechanism

Before starting with the investigation carefully check the mechanical parts of the welding equipment.
Pay particular attention to the following:

- Supply of parts to be welded;
- Positioning;
- Increase and decrease in welding force;
- Synchronised functioning of machine time and the welding process. For instance welding pressure must not decrease before the welding pulse finishes, and the pulse must not start before welding force is built up.

2 Welding head

Although this belongs to the above as well, the welding head is worth mentioning. Welding force needs to be built up gradually. Weld resistance is influenced unfavourably when large forces occur, in particular in projection welding. Vibration which allows variation in welding force must be damped.

- Is the welding force adjustment correct? Check by measuring.
- Is the movement without friction?
- Is the mass of the movable part low enough?

See also section 9.

3 Connections

Check for correct connection of the welding power supply to the mains. DC equipment must be connected to a three phase supply.

Phase 1	R	Brown
Phase 2	S	Blue
Phase 3	T	Black
Earth		Yellow/green

In DC units there is no neutral terminal as it is made using delta connection.

4 Overloading DC power equipment

DC power supply units are equipped with current stabilisation, which is normally used. One should ensure that the resistance of secondary cables does not exceed a maximum limit. If it does, this is recognizable by the current pulse waveform, the current is not equal to the adjusted value. When, for instance, the secondary transformer voltage is adjusted to e_s = 4 V and the weld resistance R_1 = 1 m, standard equipment of i_o = 1800 A can supply this current only when the resistance of the current cables does not exceed the value of $R_s = \frac{4}{1800} - \frac{1}{1000} =$ 1.22 mΩ. When e_s = 6 V then the maximum allowed value of R_s = 2.33 mΩ.

5 e_s = secondary voltage 4 or 6 V

When e_s = 6 V is available from a DC unit and less than 4 V is used, 2it Ws of extra energy must be dissipated. Use an operating voltage that corresponds as closely as possible to the transformer rating.

6 Duty cycle

Most DC units have for maximum i_o an allowed duty cycle of 5%. This means that the unit can supply maximum current for 50 ms per second. For other current values, i, the allowed duty cycle

$$= \left(\frac{i_{max}}{i}\right)^2 50 \text{ ms.}$$

For instance with a standard unit of i_{max} = 1800 A with i = 1000 A the allowed duty cycle $= \left(\frac{1800}{1000}\right)^2 50 =$ 162 ms.

When these conditions are not exceeded the thermal protection will temporarily switch off the machine.

7 Measurement

Measuring equipment is essential in failure analysis, principally for measuring current, voltage, movement and welding force, see section 8. Pushing in the 'inhibit' switch of the DC equipment permits monitoring of the control unit as if the unit is working but without supplying power. Remember to measure with earth disconnected.

8 Magnetism

Surrounding parts near to those which carry welding current must be made of non-magnetic material to prevent attraction of magnetic material. This is especially important when using DC equipment.

9 Electrical circuit

To know a part of the e-i characteristic, it is advisable to determine the complete e-i characteristic from the 50 Hz transformer which is used for welding and find out how variations in weld resistance R_1 react, see section 4.

10 Welding with parallel electrodes

Try, as far as possible to make a single weld spot per secondary welding circuit. When there are two welding positions in series, Fig. 15.1, this ensures the same current is flowing through both welding spots, and hence the required voltage is then twice as much.

When using the set up shown in Fig. 15.2. you have to take into account that the current passage of both welding spots is not the same. This will be complicated when two welds have to be placed next to each other, according to Fig. 15.3, when procedure shunt

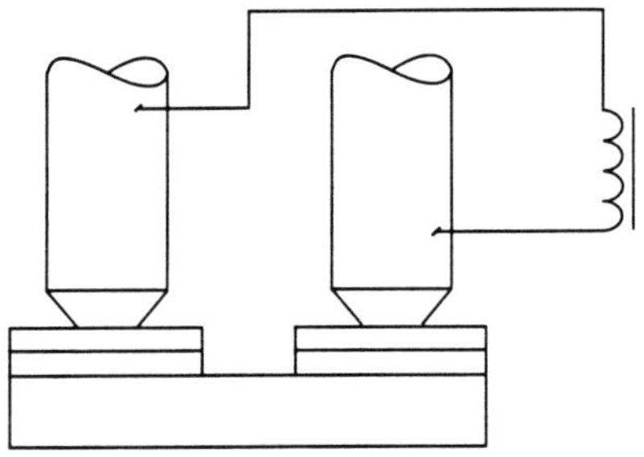

15.1 Series welding.

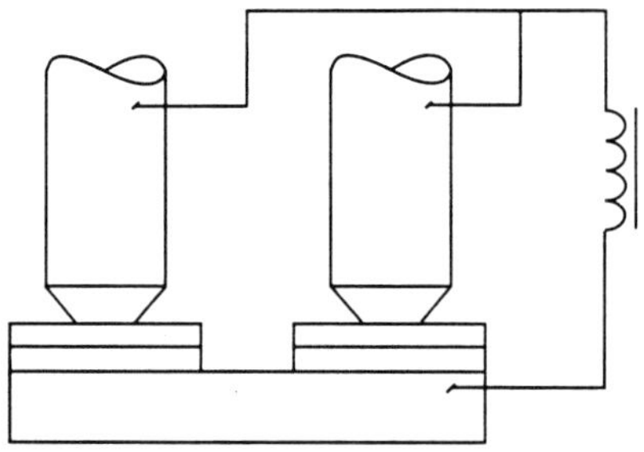

15.2 Parallel welding.

effects are involved. In this case the set-up shown in Fig. 15.4. is recommended, even though shunt effects are also possible. The last two examples occurred when welding brackets to transformers.

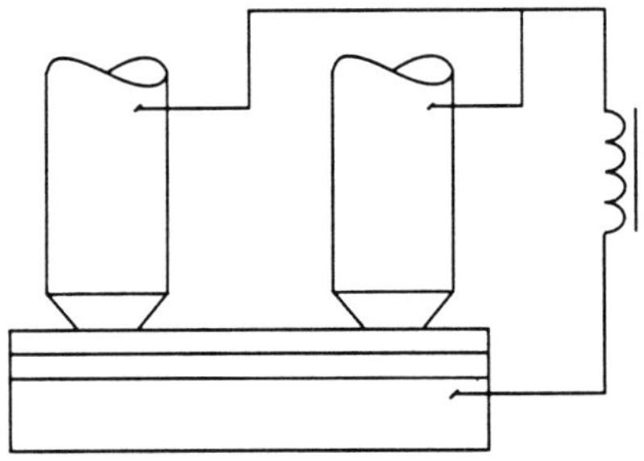

15.3 Shunt effects.

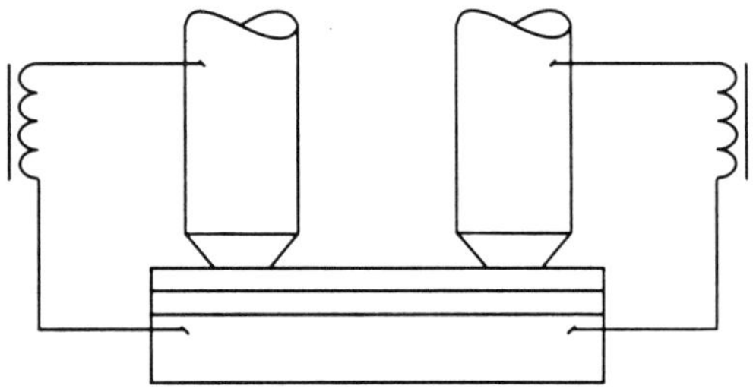

15.4 Two welding circuits.

FURTHER READING

6 Contact resistance

1 Ragnar Holm, 'Electrical contacts, theory and applications' Springer Verlag, Berlin/New York, 1967.

2 J G Kaiser, G J Dunn, T W Eager, 'The effect of electrical resistance on nugget formation during spot welding'. Welding Research Suppl, Welding Journal, June 1982, p 167S-174S.

7 Electrodes

1 L D Connell, 'Electrodes for resistance welding'. Metal Construction, Jan 1977, p 30-44.

2 R M Rivett, S A Westgate, 'Resistance welding of aluminium alloys in mass production'. Metal Construction, Oct 1980, p 510-517.

3 S L Roswell, 'Resistance spot welding galvanised steel sheet, electrode life and corrosion tests'. Research Report 17/1976/P, Sept 1976, The Welding Institute.

4 Wearing of electrodes in welding of tinplated and unalloyed steel, 'Schweisstechnischen Lehr- und Versuchsanstalt München'. Schweissen und Schneiden, 33 (1980) 1.

5 K Rust, 'CuCr-Ti - a new resistance weld electrode material'. 8 ZIS-Mitteilungen 8/75, p 1488-1492.

9 Welding dynamics

1 M P Koster, 'Predicting the dynamic behaviour of cam mechanisms'. Philips training course for designers in production mechanisation and product design, August 1974.

2 J Ruge, B Kenmann, 'Forces in resistance welding'. Schweissen und Schneiden, 33 p 170s.

3 D W Dickinson, J E Franklin, A Stanya, 'Characterisation of spot welding behaviour of dynamic electrical parameter monitoring'. Welding Journal, June 1980, p 170S.

4 V G Novokreshchenov, 'Automatic adjustment of the initial compression force in micro spot welding'. Welding Production, 19 (1972)9, p 33-35.

10 Thermal management

1 M Bogman, Study report, Technical Highschool, 1980.

2 J O Zwolsman, 'The thermal management of the resistance spot welding process'. CFT Report 12/84E.

3 J O Zwolsman, 'Temperature calculations in resistance welding'. DVS-Berichte, Band 70, September 1981.

11 Weld splashes

1 K I Johnson, M D Hannah, S L Roswell, W O Dinsdale, 'Properties of splashed resistance spot welds'. Metal Construction and British Welding Journal, Nov 1973, p401-406.

2 K I Johnson, M D Hannah, S L Roswell, W O Dinsdale, 'Characterisation of splashed resistance spot welds in mild steel sheet'. The Welding Institute, Research Report P/49/72.

3 G A Mirlin, V S Savchenko, V S Schedrow, 'Formation of cracks in spot welding and methods of prevention'. BWRA, Welding Production, May 1964.

4 K C Wu, 'The mechanism of expulsion in weld bonding of anodized aluminium'. Welding Research Suppl, Welding Journal, August 1977, p 123S.

5 AWS Welding Handbook, 'Projection welding'. Part II, 1969, p 26-41.

12 Seebeck-, Peltier- and Thomson-effect

1 T Balder, 'The influence of the Peltier effect in resistance welding'. De Ingenieur, 74 (1962) 48, 30 November. Also in: Philips Technisch Tijdschrift, volume 20, p 174-178.

2 J J Lander, 'Measurements of Thomson coefficients for metals at high temperatures and of Peltier coefficients for

solid/liquid interfaces of metals'. Physical Review, 74 (1948) 4, August 15.

3 M Hasir, 'Investigation on the Peltier effect during resistance welding of tin-plated steel with Tungsten insert electrodes,' Schweissen und Shneiden, 36 (1984) p116-121.

4 Publication 'Physical properties of materials', p 117-122. Further information is unknown.

5 H D Wendler, Scheel, Pause, 'The influence of the Peltier effect to the shape of the welding lens in resistance welding'. Message from Institute fur Schweisstechnik Technical University 'Otto von Guericke', Magdeburg 9 ZIS Mitteilungen, 3/68 p 446-457.

13 Welding equipment

1 L Schuringa, J O Zwolsman, 'Resistance welding of small components with transistor controlled DC power sources'. DVS Lecture, Essen, 7-8 May 1985.

14 Process control

1 A P Hulst, JO Zwolsman, 'Adaptive control in resistance welding of small components'. Polytechnisch Tijdschrift W, 1976, Nr. 6-383.

2 J C Needham, 'In-process quality control and automation in welding'. Part 1, 'General considerations'. Research Bulletin, The Welding Institute, March 1972. Part 2, 'Application of basic principles'. Research Bulletin, The Welding Institute, April 1972.

3 D R Andrews, D M K Walls, 'Dynamic resistance monitoring applied to projection welding'. Metal Construction, Jan 1980, p 30-33.

4 D R Andrews, J Broomhead, 'Quality assurance for resistance spot welding'. Welding Journal, June 1975, p 431-435.